Nadja Grobe

Biosynthesis of Morphine in Mammals

Nadja Grobe

Biosynthesis of Morphine in Mammals

Südwestdeutscher Verlag für Hochschulschriften

Impressum / Imprint
Bibliografische Information der Deutschen Nationalbibliothek: Die Deutsche Nationalbibliothek verzeichnet diese Publikation in der Deutschen Nationalbibliografie; detaillierte bibliografische Daten sind im Internet über http://dnb.d-nb.de abrufbar.
Alle in diesem Buch genannten Marken und Produktnamen unterliegen warenzeichen-, marken- oder patentrechtlichem Schutz bzw. sind Warenzeichen oder eingetragene Warenzeichen der jeweiligen Inhaber. Die Wiedergabe von Marken, Produktnamen, Gebrauchsnamen, Handelsnamen, Warenbezeichnungen u.s.w. in diesem Werk berechtigt auch ohne besondere Kennzeichnung nicht zu der Annahme, dass solche Namen im Sinne der Warenzeichen- und Markenschutzgesetzgebung als frei zu betrachten wären und daher von jedermann benutzt werden dürften.

Bibliographic information published by the Deutsche Nationalbibliothek: The Deutsche Nationalbibliothek lists this publication in the Deutsche Nationalbibliografie; detailed bibliographic data are available in the Internet at http://dnb.d-nb.de.
Any brand names and product names mentioned in this book are subject to trademark, brand or patent protection and are trademarks or registered trademarks of their respective holders. The use of brand names, product names, common names, trade names, product descriptions etc. even without a particular marking in this works is in no way to be construed to mean that such names may be regarded as unrestricted in respect of trademark and brand protection legislation and could thus be used by anyone.

Coverbild / Cover image: www.ingimage.com

Verlag / Publisher:
Südwestdeutscher Verlag für Hochschulschriften
ist ein Imprint der / is a trademark of
AV Akademikerverlag GmbH & Co. KG
Heinrich-Böcking-Str. 6-8, 66121 Saarbrücken, Deutschland / Germany
Email: info@svh-verlag.de

Herstellung: siehe letzte Seite /
Printed at: see last page
ISBN: 978-3-8381-3605-9

Zugl. / Approved by: Halle, Martin-Luther-Universitaet Halle-Wittenberg, Diss., 2009

Copyright © 2012 AV Akademikerverlag GmbH & Co. KG
Alle Rechte vorbehalten. / All rights reserved. Saarbrücken 2012

Table of Contents

I	INTRODUCTION	4
II	MATERIAL & METHODS	13
1	Animal Tissue	13
2	Chemicals and Enzymes	13
3	Bacteria and Vectors	13
4	Instruments	14
5	Synthesis	15
5.1	Preparation of DOPAL from Epinephrine (according to DUNCAN 1975)	15
5.2	Synthesis of (R)-Norlaudanosoline*HBr	15
5.3	Synthesis of [7D]-Salutaridinol and [7D]-epi-Salutaridinol	16
6	Application Experiments	18
6.1	Application to Papaver setigerum Seedlings	18
6.2	Injections into Mice and Collection of Urine	18
7	Preparation of Protein	18
7.1	Preparation of Crude Liver Protein	18
7.2	Preparation of Microsomes from Liver	18
7.3	Heterologous Expression in E.coli	19
7.3.1	Cloning of Amine N-Methyltransferase cDNA in Vector pET28a and Transformation into E. coli PlusKJ	19
7.3.2	Isolation of Amine N-Methyltransferase His-tagged Protein from E. coli	21
7.3.3	Thrombin Digestion of His-tagged Protein	22
8	Standard Test for Mammalian Liver Enzymes	22
9	Solid Phase Extraction of Alkaloids	23
9.1	Strata X-C (Phenomenex)-TCA method	23
9.2	Strata X-C (Phenomenex)-HClO$_4$ method	23
9.3	Bond Elut Certify (Varian)	23
9.4	Sep Pak Plus C18 (Waters)-basic conditions	24
10	Chromatographic methods	24
10.1	Thin Layer Chromatography (TLC)	24
10.2	Analytical High Performance Liquid Chromatography	24
10.3	Preparative High Performance Liquid Chromatography	25
10.4	Liquid Chromatography-Mass Spectrometry (LC-MS)	25
10.5	Liquid Chromatography-High Resolution Mass Spectrometry (HR-LC-MS)	27
11	Protein Assay	28
12	Radioactivity Measurements	28
III	RESULTS	29
PART A: MORPHINE ANALYTICS		29
1	Stability of morphine	29
2	Solid phase extraction with Strata X-C	32
3	Ion suppression	34
4	MS-Instruments and Sensitivity	38

PART B:	I.P. INJECTIONS INTO MICE AND THE MORPHINE PATHWAY IN MAMMALS		41
1	The Biogenesis of Mammalian Morphinan Alkaloids		41
1.1	Intraperitoneal Injection of (R,S)-Norlaudanosoline into Mice		44
1.1.1	Urinary Excretion of Tetrahydrobenzylisoquinoline Alkaloids after Application of (R,S)-Norlaudanosoline		47
1.1.2	Urinary Excretion of Tetrahydroprotoberberine Alkaloids after Application of (R,S)-Norlaudanosoline		48
1.1.3	Urinary Excretion of Aporphine and Morphinan Alkaloids after Application of (R,S)-Norlaudanosoline		49
1.2	Intraperitoneal Injection of (R)-Norlaudanosoline into Mice		50
1.3	Intraperitoneal Injection of (R)-[N-CD$_3$]-Reticuline into Mice		52
2	Thebaine formation from Salutaridine in Mammals		56
2.1	Intraperitoneal Injection of Salutaridine into Mice		57
2.2	Intraperitoneal Injection of [7D]-Salutaridinol into Mice		59
3	Morphine formation from Thebaine in Mammals		62
3.1	Intraperitoneal Injection of [N-CD$_3$]-Thebaine into Mice		64
3.2	Intraperitoneal Injection of Oripavine into Mice		65
PART C:	ENZYME STUDIES IN SELECTED MORPHINE BIOSYNTHETIC STEPS		69
1	Formation of 3,4-Dihydroxyphenylacetaldehyde (DOPAL) in Animals		70
2	Condensation to Norlaudanosoline		74
3	N-Methylation in Mammalian Morphine Biosynthesis		78
3.1	Heterologous Expression of two Human N-Methyltransferases		79
3.2	Activity of Recombinant Phenylethanolamine N-Methyltransferase and Amine N-Methyltransferase		81
3.3	Determination of K_m, V_{max} and the Catalytic Efficiency of NMT with (R)-configurated THBIQ Alkaloids as Substrates		83
4	The Oxidative C-C-Phenol-Coupling in Mammals		87
4.1	The Phenol-Coupling Reaction in Mammals Catalyzed by Human CYP 2D6 and CYP 3A4		88
4.2	The Phenol-Coupling Reaction in Mammals Catalyzed by Rat CYP 2D2		92
4.3	Kinetic Analysis and Characteristics of Mammalian P450 Enzymes Catalyzing the Phenol Coupling Reaction		95
5	3-O-Demethylation of Thebaine and Codeine in Mammals Catalyzed by Human CYP 2D6 and Rat CYP 2D2		100
6	6-O-Demethylation of Thebaine and Oripavine		104
IV	DISCUSSION		106
V	SUMMARY		122
VI	REFERENCES		125

List of Abbreviations

APS	ammonium persulfate
cDNA	complementary deoxyribonucleic acid
cps	counts per second
CPR	cytochrome P450 reductase
CYP	cytochrome P450
DLPC	L-α-dilauroyl-sn-glycero-3-phosphocholine
DNA	deoxyribonucleic acid
DOPAL	3,4-dihydroxyphenylacetaldehyde
DOPA-pyruvate	3,4-dihydroxyphenylpyruvate
DOPET	3,4-dihydroxyphenylethanol
DW	dry weight
E. coli	*Escherichia coli*
EC-HPLC	high perfomance liquid chromatography coupled to electrochemical detection
EDTA	ethylenediaminetetraacetic acid
ELISA	enzyme-linked immunosorbent assay
EPI	enhanced product ion
EtOH	ethanol
FW	fresh weight
GC-MS	gas chromatography-mass spectrometry
GDP	guanosine diphosphate
GTP	guanosine triphosphate
HPLC	high perfomance liquid chromatography
HR-LC-MS	liquid chromatography-high resolution mass spectrometry
HR-MS	high resolution mass spectrometry
i.p.	intraperitoneal
LC-MS	liquid chromatography-mass spectrometry
LOD	limit of detection
LOQ	limit of quantitation
MRM	multiple reaction monitoring
MS	mass spectrometry
NAD(H)	nicotinamide adenine dinucleotide
NADP(H)	nicotinamide adenine dinucleotide phosphate
NMR	nuclear magnetic resonance
OD_{600}	optical density at 600 nm
P450	cytochrome P450
PCI	phenol-chloroform-isoamylalcohol
PCR	poymerase chain reaction
RIA	radioimunnoassay
rpm	round(s) per minute
SAM	*S*-adenosyl-L-methionine
SDS	sodium dodecyl sulfate
SDS-PAGE	sodium dodecyl sulfate polyacrylamide gel electrophoresis
SPE	solid phase extraction
TCA	trichloroacetic acid
TFA	trifluoroacetic acid
THBIQ	tetrahydrobenzylisoquinoline
ThDP	thiamine diphosphate
THPB	tetrahydroprotoberberine
TIQ	tetrahydroisoquinoline
TLC	thin layer chromatographgy
Tris	tris(hydroxymethyl)aminomethane
UV	ultraviolett
vol	column volume

I Introduction

Morphine, one of the strongest analgesics known, is derived from opium poppy (*Papaver somniferum*), a Mediterranean plant that is considered to be the oldest cultivated medicinal plant in Europe. Morphine occurs among *ca.* 80 other alkaloids in the latex of opium poppy in substantial amounts of up to 30%. The correct chemical structure of morphine was first proposed in 1925 by GULLAND & ROBINSON and later confirmed by SCHÖPF 1927 (degradation studies), GATES & TSCHUDI 1952, 1956 (total synthesis in a 20 step procedure) and MACKAY & HODGKIN 1955 (X-ray crystallography). Morphine was first isolated from opium more than 200 years ago by the pharmacist's assistant Friedrich W.A. Sertürner (SERTÜRNER 1806, 1817). Because of its characteristic to put people to sleep Sertürner named the bitter tasting compound "morphinum" after Morpheus, the Greek god of dreams. Sertürner's groundbreaking discovery of the biologically active, principal substance of opium (that possessed 10 times the power of processed opium) led not only to a change in the field of drug therapy but also made him the founder of a new class of nitrogen-containing compounds, i.e. alkaloids. The name "alkaloid" for this new class was coined by MEISSNER (1819), an apothecary from Halle, Germany, introducing these types of molecules simply as plant derived substances that act like alkali. This definition was further expanded in structural terms by WINTERSTEIN & TRIER (1910) describing alkaloids as compounds of plant origin with heterocyclic-bound, basic nitrogen that have a complex molecular structure and pharmacological activity, a definition that is still valid today.

Morphine belongs to the benzylisoquinoline type of alkaloid with two additional ring closures. Its nitrogen-carbon skeleton consists of a total of five condensed rings and contains five asymmetric centers (Fig. 1). The tertiary amino group of the N-methylpiperidine ring accounts for the basic characteristic of morphine. Because of its stereochemical complexity, to date a commercial synthesis with a competitive price has not been achieved. Being prescribed for medical purposes including relief in severe chronic pain from e.g. cancer, advanced medical illness or post-operative analgesia, more than 200 tons of morphine are provided per year to satisfy the market. Licit production and sales of morphine generate 5 billion dollars annually at the pharmacy level worldwide and 50-100 billion dollars are generated from illicit sales. The dried milk juice obtained from opium poppy is the principal commercial source for the alkaloid. A biotechnological approach to produce morphine is anticipated but will only be feasible if the key enzymes of the plant biosynthesis and their regulations are known.

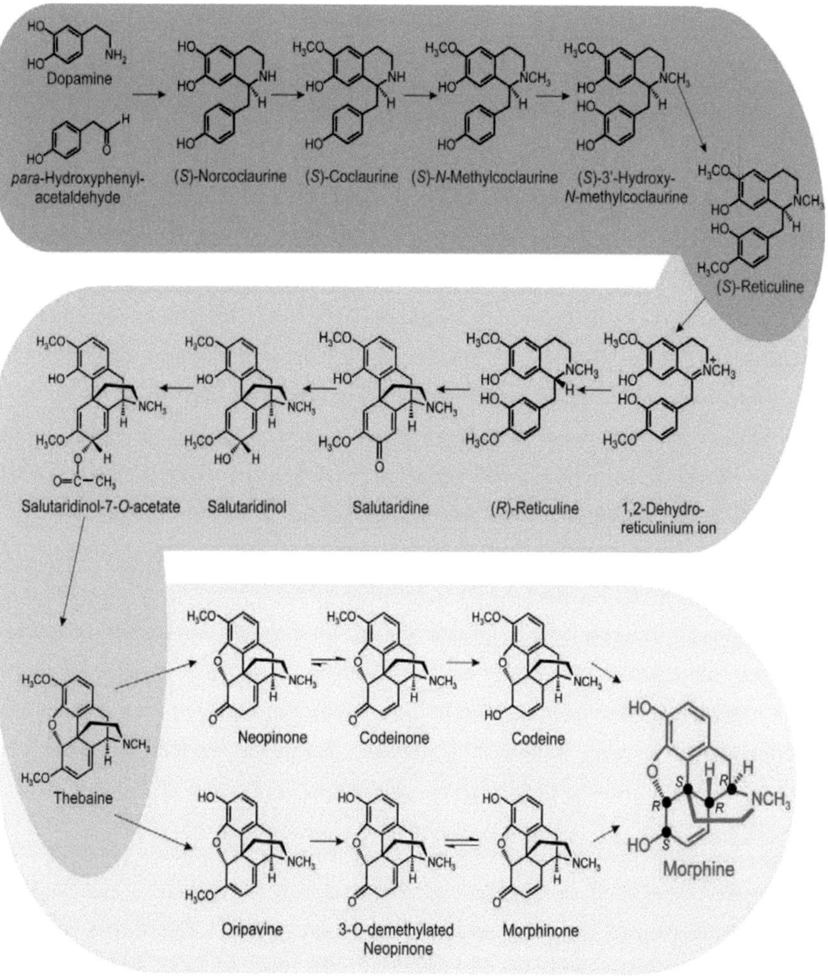

Fig. 1: The biosynthesis in opium poppy starts with the condensation of the two catechols, dopamine and *para*-hydroxyphenylacetaldehyde, to (S)-norcoclaurine that is further converted *via* two additional ring closures to morphine (for review KUTCHAN 1998). The five asymmetric centers in the morphine molecule are shown in black.

The biosynthesis of morphine in opium poppy has been completely elucidated in the past 25 years mainly by the groups of M.H. ZENK and T.M. KUTCHAN. A scheme of the biosynthetic pathway starting from dopamine and *para*-hydroxyphenylacetaldehyde to morphine is shown in Fig. 1 (KUTCHAN 1998). Most of the enzymes that are involved in the 17 enzymatically catalyzed steps have been isolated and functionally characterized; most of them are cloned. The pathway can be divided into three steps: 1) production of the benzylisoquinoline alkaloid and central intermediate (S)-reticuline, 2) conversion of (S)-reticuline into (R)-reticuline to build the pentacyclic morphinan precursor thebaine, and 3) the entering of thebaine into the bifurcate pathway to produce morphine *via* codeine or oripavine.

This plant derived alkaloid morphine exhibits besides its analgetic effects several other pharmacological effects like euphoria, withdrawal, respiratory depression, constipation, psychiatric syndromes and immunosuppression which are believed to be mediated through the µ-opioid receptor. The receptor is a seven transmembrane protein that is associated with a heterotrimeric G-protein consisting of an α-, β- and γ-subunit. A classical G-protein mediated signal cascade is activated after binding of morphine to the receptor. The interaction leads to a conformational change of the µ-receptor that allows the exchange of GDP with GTP bound to the α-subunit of the G-protein and immediate dissociation into its subunits. The dissociated α-GTP subunit activates an adenylylcyclase that further triggers several mechanisms resulting in the pharmacological effects of morphine. Hydrolysis of GTP to GDP by the hydrolase activity of the α-subunit allows reassociation with the β- and γ-subunit and thus a restart of the cycle. Morphine is prescribed as one of the most powerful pain-relieving compounds known; however, the development of tolerance during chronic pain treatment is a severe obstacle in clinical usage. Studies suggested that the inability of morphine to cause internalization of the µ-receptor upon binding hinders the receptor from entering into the recycling pathway. That results in desensitized receptors remaining on the plasma membrane which could be a reason for morphine tolerance (KEITH *et al.* 1996, ZHANG *et al.* 1998, SCHULZ *et al.* 2004).

Because of its physiological effects in humans and binding activity to a specific receptor the idea of an endogenous occurrence of morphine in animals was formed. The French researcher MAVROJANNIS was the first one who suggested in 1903 that morphine could be present endogenously in animals based on his investigations in rats suffering catalepsy, a disease of the central nervous system (MAVROJANNIS 1903). He observed that the exact same symptoms of the disease were induced in control rats after application of morphine. Years later, the group of SPECTOR (Roche Institute, New Jersey, USA) reported for the first time a positive result with the isolation of a morphine-like substance from brain of various animals by immunological methods (GINTZLER *et al.* 1976a). This morphine-like compound was described to recognize and bind opioid receptors of a mouse neuroblastoma-glioma hybrid cell line (BLUME *et al.* 1977). Similarly, the presence of the non-peptide morphine-like compound was verified in human cerebrospinal fluid (SHORR *et al.* 1978), mouse brain (GINTZLER *et al.* 1978) as well as in calf brain (KILLIAN *et al.* 1981). With the goal to identify the morphine-like opioid, OKA *et al.* (1985) reported the purification of the compound from toad skin of *Bufo marinus* (800 g toad skin from 50 toads) and found it to be identical to morphine by immunological, pharmacological and physical chemical criteria.

The group of GOLDSTEIN (Stanford University, California, USA) and SPECTOR reported in the following year numerous times on the occurrence of morphine in animals such as in brain and adrenals of beef (GOLDSTEIN *et al.* 1985, WEITZ *et al.* 1986), in rat brain (DONNERER *et al.* 1986, WEITZ et al. 1986) as well as in 24-hr urine of rats (DONNERER *et al.* 1987) (Tab. 1). The claim of both groups that the detected morphine was of endogenous origin raised a controversy since morphine was also found in various natural products such as milk, hay and lettuce (HAZUM *et al.* 1981) as well as in rodent food pellets (ROWELL *et al.* 1982) suggesting a dietary origin of morphine in animal organs (Tab. 1). This hypothesis of exogenously introduced morphine in mammals could not be refuted by the researchers around Goldstein and Spector since they had no direct evidence that their detected mammalian morphine was of endogenous origin. A former student and follower of Spector, G.B. STEFANO, strongly supported the hypothesis that morphine in animals is of endogenous origin. Several articles published by Stefano's group in the past 15 years attempted to demonstrate the occurrence of endogenous morphine e.g in human plasma (LIU *et al.* 1997), rat adrenal gland (GOUMON & STEFANO 2000), PC-12 cells, an adrenal medullary chromaffin cell line derived from rat pheochromocytoma (GOUMON *et al.* 2000a), rat brain (GOUMON *et al.* 2000b), marine mollusks *Mytilus edulis* and *Modiolus deminissus* (ZHU *et al.* 2001a, GOUMON *et al.* 2001), human heart tissue (ZHU *et al.* 2001b) and mouse brain (NERI *et al.* 2008) (Tab. 1). Again, these results of the Stefano group lacked the clear evidence that the detected morphine was of endogenous origin since it could not be excluded that the alkaloid could have also been exogenously introduced through e.g. contamination and diet. Moreover, quantitation of morphine was done by HPLC coupled to electrochemical detection (EC-HPLC), that was described to be prone to several pitfalls such as possible contamination of the electrodes or changes in the flow conditions or in the environment of the electrode or the electrochemical cell itself. EC-HPLC could therefore not only yield false positive results but also lacks, unlike NMR and MS, the ability to unequivocally identify the compound of interest and to discriminate it against background interferences.

The reports of the group around M.H. ZENK dispelled for the first time the discussion on the existence of an endogenous synthesis of morphine in humans by demonstrating that human neuroblastoma cells are capable of producing the alkaloid morphine found to be present in 10 nM concentration endogenously in these cells (POEAKNAPO *et al.* 2004, BOETTCHER *et al.* 2005) (Tab. 1).

Tab. 1: Occurrence of morphine in animals.

Author	Year	Occurrence	Amount	Quantitation
HAZUM et al.	1981	Cow and human milk	702-1754 pmol/l (200-500 ng/l)*	HPLC
		Hay and lettuce	7-35 pmol/g DW (2-10 ng/g DW)*	
ROWELL et al.	1983	Rodent food pellets	18-53 pmol/g DW (5-15 ng/g DW)*	RIA
OKA et al.	1985	Toad skin	3 pmol/g FW	RIA
GOLDSTEIN et al.	1985	Bovine		RIA
		Hypothalamus	≥4.9 pmol/g FW	
		Adrenal	4500-33000 pmol/g FW	
DONNERER et al.	1986	Rat:		RIA
		Brain	0.026 pmol/g FW	
		Intestines	0.017 pmol/g FW	
		Liver	0.011 pmol/g FW	
		Kidney	0.016 pmol/g FW	
		Blood	0.002 pmol/g FW	
WEITZ et al.	1986	Bovine hypothalamus	5-25 pmol/g FW	RIA
		Bovine adrenal	≥100 fmol/g FW	
		Rat brain	≥200 fmol/g FW	
DONNERER et al.	1987	Rat:		RIA
		Brain	2.85 pmol/g FW	
		Spinal cord	3.1 pmol/g FW	
		Heart	0.7 pmol/g FW	
		Adrenals	6.3 pmol/g FW	
		Liver	0.07 pmol/g FW	
		Submandibular gland	0.07 pmol/g FW	
		Skin	0.26 pmol/g FW	
		Urine	0.233 pmol/day	
MATSUBARA et al.	1992	Human urine	2.93 pmol/ml	GC-MS
MIKUS et al.	1994	Human urine	0.02-27 pmol/ml (0.005-7.6 ng/ml)*	RIA
LIU et al.	1997	Human plasma	0.28 pmol/ml (80 pg/ml)*	EC-HPLC
GUARNA et al.	1998	Rat brain	0.7 pmol/g FW(0.2 ng/g FW)*	GC-MS
HOFMANN et al.	1999	Human urine	0.21 pmol/ml (310 pmol/ 1.5 l)*	GC-MS
GOUMON & STEFANO	2000	Rat adrenal gland	368 pmol/ g FW (105 ng/g FW)*	EC-HPLC
GOUMON et al.	2000a	PC-12	6 pmol/million cells (1.7 ng/million cells)*	EC-HPLC
GOUMON et al.	2000b	Rat brain	25 pmol/g FW (7 ng/g FW)*	EC-HPLC
GOUMON et al.	2000c	*Ascaris suum*	4098 pmol/g FW (1168 ng/g FW)	EC-HPLC
ZHU et al.	2001a	*Mytilus edulis*	9.4 pmol/ganglia (2.67 ng/ganglia)*	EC-HPLC
ZHU et al.	2001b	human heart tissue	372 pmol/g FW (106 ng/g FW)*	EC-HPLC
GOUMON et al.	2001	*Modiolus deminissus*	8.5 pmol/ganglia (2.41 ng/g/g FW)*	EC-HPLC
GUARNA et al.	2002	Mouse brain	1 pmol/g FW (0.3 ng/g FW)*	GC-MS
POEKNAPO et al.	2004	SH-SY5Y	15 fmol/million cells	GC-MS
ZHU et al.	2005	Human:		EC-HPLC
		White blood cells	0.04 pmol/million cells (12.33 pg/million cells)*	
		Polymorphonuclear cells	0.04 pmol/million cells (11.2 pg/million cells)*	
BOETTCHER et al.	2006	Human:		GC-MS
		Polymorphonuclear cells	0.04 pmol/million cells (10.4 pg/million cells)*	
		Mononuclear cells	0.03 pmol/million cells (8.5 pg/million cells)*	
		Erythrocyte fraction	0.28 pmol/ml packed volume (81 pg/ml packed volume)*	
NERI et al.	2008	Mouse brain	0.5 pmol/g FW (140 pg/g FW)	GC-MS
MULLER et al.	2008	Mouse brain	16 pmol/g FW	ELISA

* values found in literature, converted into pmol.

The researchers showed by GC- MS analysis that $^{18}O_2$ was incorporated into morphine when human neuroblastoma cells were grown in a closed system under sterile conditions in $^{18}O_2$ atmosphere (POEAKNAPO et al. 2004). Additionally, feeding with heavy-isotope labeled distant precursors of morphine biosynthesis to human neuroblastoma cells led to position-specific labeling of endogenous morphine, as established by GC-MS/MS (BOETTCHER et al. 2005). Based on these results a biosynthetic pathway of morphine in humans was predicted starting from L-tyrosine (Fig. 2).

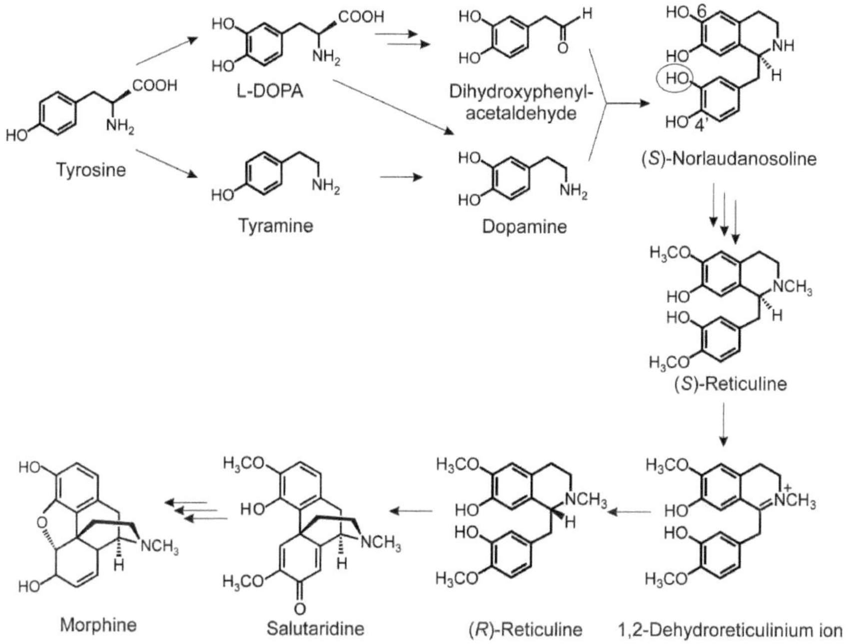

Fig. 2: The biosynthesis of morphine in humans as proposed by BOETTCHER et al. (2005). The condensation of the two catechols, dopamine and dihydroxyphenylacetaldehyde yields (S)-norlaudanosoline, the first alkaloid precursor in the mammalian biosynthesis that differs from the trihydroxylated (S)-norcoclaurine, the precursor in the plant pathway, in one additional hydroxy group (circled in black). Terminal steps were proposed to be identical to the plant pathway.

The proposed morphine biosynthesis in humans revealed a fundamental difference to the opium poppy pathway in such that the mammalian morphine was postulated to originate from the tetrahydroxylated tetrahydrobenzylisoquinoline (THBIQ) alkaloid norlaudanosoline rather than the trihydroxylated THBIQ alkaloid norcoclaurine, the alkaloid precursor in plants.

A natural occurrence of norlaudanosoline in animals has been reported previously numerous times. The tetrahydroxylated THBIQ alkaloid was detected in rat brain (CASHAW 1993a, HABER et al. 1997, SÄLLSTRÖM BAUM et al. 1999), in human brain (SANGO et al. 2000) as well as in urine of humans (SANDLER et al. 1973, MATSUBARA et al. 1992). The two alkaloids codeine and morphine were also naturally detected in human urine of 40 subjects by RIA analysis revealing highly individual variations in the concentration of codeine ranging from 0.003 pmol/ml urine to 7 pmol/ml urine and morphine ranging from 0.02 pmol/ml urine to 27 pmol/ml urine (more than 1000-fold difference) (MIKUS et al. 1994). The same group established a few years later the "gold standard" for the detection of morphine in human urine by an exceptionally sensitive GC-MS/MS method that allowed the measurement of morphine in urine samples down to a concentration of 2.5 fmol/ml urine (HOFMANN et al. 1999). The study showed that amounts of excreted morphine decreased drastically from ca. 500 fmol/ml urine on the first day under normal diet to 25 fmol/ml urine five days after substitution with the morphine-free liquid diet. The urinary excretion of morphine by humans was also shown very recently by the group of SPITELLER with a very sensitive analytical LC-MS/MS method. With a limit of detection of 35 fmol morphine/ml the presence of the alkaloid in 28 dansyl-derivatized urine samples was clearly confirmed (LAMSHÖFT & SPITELLER 2009, in preparation). The observed high individual variations of 0.01 - 329 pmol morphine/ml urine were very similar to the ones reported by MIKUS et al. (1994). A biosynthetical link between the biomolecules L-DOPA, norlaudanosoline, codeine and morphine in animals was suggested since urine of Parkinsonian patients undergoing an L-DOPA therapy (0.25-1 g L-DOPA per day) showed elevated levels of norlaudanosoline (CASHAW 1993b) as well as increased levels of the three alkaloids norlaudanosoline, codeine and morphine (MATSUBARA et al. 1992). This was an intriguing hypothesis that lacked, however, the clear conclusive evidence that the detected alkaloids really originated from the administered L-DOPA.

The presence of morphine and potential precursors in body fluids and tissues of animals and humans which has been published to date is an indication but surely no proof of an endogenous origin of these alkaloids in animals. Because these studies did not conclusively distinguish between endogenous and exogenously introduced (dietary) alkaloids the hypothesis of endogenously synthesized morphine continued to be questioned. Clear evidence for an endogenous synthesis of the morphinan and morphine alkaloids in mammals can only come by administration of heavy-isotope labeled, biosynthetic precursors to the animals and the detection of heavy-isotope labeled morphinan and morphine alkaloids formed *in vivo* from these precursors. This idea was used by POEAKNAPO et al. (2004) and BOETTCHER et al. (2005) and revealed a biosynthesis of morphine in human neuroblastoma cells; however, force feeding of potential precursors to human cells outside of the body with relaxed, promiscuous enzymes could possibly have led to an ambiguous morphine synthesis.

One goal of this thesis was to address this criticism and to establish a different biochemical strategy to show that living animals are capable of biosynthesizing morphine *de novo*. A direct involvement into the mechanism and biosynthesis of morphine in living animals could possibly be achieved by intraperitoneal (i.p.) injection into mice of potential precursors that preferably carry a heavy-isotope label. The question was whether metabolites of these administered biosynthetic precursors would be excreted in the urine of mice. While the occurrence of human alkaloids in mouse and rat tissue has been described by several researchers in the past 20 years (Tab. 1) the urinary excretion of morphine by rodents has so far only been reported on one occasion (DONNERER et al. 1987). Would morphine alkaloids be detectable in the urine of the i.p. injected mice and carry position-specific isotopic labeling according to the heavy-isotope label of the i.p. injected precursor thus providing evidence for an endogenous synthesis? In an early hypothesis of DAVIS & WALSH (1970), norlaudanosoline, that was later found in animal tissue and body fluids numerous times, was proposed to be converted in animals into more complex alkaloids such as morphine but the researchers did not give experimental evidence for their hypothesis. Norlaudanosoline was therefore chosen as the first potential biosynthetic precursor for i.p. injection into mice. If the three segments of morphine biosynthesis, similar as indicated in the opium poppy pathway in Fig. 1, could be also shown in the living animal with the i.p. injection of preferably heavy-isotope labeled precursors, conclusive evidence for an endogenous origin of morphine in mammals would be provided. The search for urinary metabolites was therefore based on the plant morphine pathway (Fig. 1) as well as on the proposed pathway of morphine in humans (Fig. 2). The idea was to search with a correct and reliable analytical method for heavy-isotope labeled metabolites in the urine of i.p. injected mice. With such an exact analytical method, metabolites in urine of i.p. injected mice that would not occur in control urine and that carry position-specific isotopic labeling could then be clearly identified. For the correct determination of urinary heavy-isotope labeled metabolites the highest sensitivity, accuracy and selectivity of a Linear Trap Quadrupole Orbitrap High-Resolution Mass Spectrometry (HR-MS) instrument was used in this thesis. HR-MS with an Orbitrap Mass Spectrometer has the clear benefit that masses as well as corresponding chemical formulae are determined with an accuracy of ≤ 2 ppm which was exploited in this thesis for the analysis of urine from i.p. injected mice. I.p. injection of heavy-istope labeled biosynthetic morphine precursors into mice combined to the state-of-the-art analytical method of HR-MS with an Orbitrap MS instrument will provide the clearness that this controversial area of research needs to unequivocally reveal if mammals are capable of synthesizing morphine endogenously.

Additional proof for this hypothesis would be given if enzymes of selected biosynthetic steps in the proposed pathway of morphine are discovered. This goal was also included within the aims of this thesis. To clearly determine product formation, enzymatic mixtures were subjected to LC-MS analysis with a 4000 Triple Quadrupole Mass Spectrometer. Despite obvious advantages of Mass Spectrometry over CE-HPLC, RIA and ELISA as analytical method to unequivocally reveal an endogenous origin of morphine in mammals, some challenges still remain. These problems were addressed in this thesis, too, such as the identification of morphine and potential precursors in animal tissue with concentrations close to the limit of detection, sample preparation procedures to purify morphinan alkaloids from biological background as well as stability of the analyte in different chemical environments.

The question if living animals are capable to biosynthesize morphine should be unequivocally given herein by taking together the findings of the HR-MS analysis of urinary metabolites of i.p. injected mice with the discovery of enzymes possibly involved in the mammalian morphine pathway.

II Material & Methods

1 Animal Tissue

Rats and mice of different genotype were obtained through VMD Ph.D. Tammie Keadle, Washington University, from Dr. Erik Herzog's lab Department of Biology at Washington University. Eight week old or retired breeders (4-8 months old) C57 black mice were purchased from Charles River (Wilmington, MA). Housing was carried out at the Animal Facility at Washington University, St. Louis. Free access to rodent chow (Teklad food pellets) and water was provided. Brain tissue of C57 black mice were also generously provided by Dr. Chang-Shen Qiu, from Dr. Robert W. Gereau's lab, Washington University Pain Center.

2 Chemicals and Enzymes

All customary chemicals and solvents were purchased from Sigma (St. Louis, MO) or Fisher Scientific (Waltham, MA). Alkaloids were from our departmental collection, and isoboldine and pallidine were gifts of Dr. André Cavé, University Paris-Sud, and Dr. Shoei-Sheng Lee, National Taiwan University. L-α-Dilauroyl-sn-glycero-3-phosphocholine, bovine liver catalase, bovine erythrocytes superoxide dismutase, horse liver alcohol dehydrogenase and catechol O-methyltransferase (porcine liver) were obtained from Sigma (St. Louis, MO). Glucose 6-phosphate dehydrogenase (yeast) was purchased from Serva (Heidelberg, Germany). Human CYP 2D6 and rat NADPH-P450 reductase were generous gifts from Dr. Fred P. Guengerich. Human CYP 2D6 and human CYP 3A4 (with or without cytochrome b_5) were purchased from BD Biosciences (Woburn, MA). Human amine N-methyltransferase was purchased as a full length clone from Deutsches Ressourcenzentrum für Genomforschung (Berlin, Germany). *E. coli* DH5α competent cells were kindly provided by Dr. Taiji Nomura. [Methyl-^{14}C]-adenosyl-L-methionine (50.43 mCi/mmol, cat. ARC 0344) and [N-methyl-^{3}H]-morphine (80 Ci/mmol, cat. ART 0659) were obtained from American Radiolabeled Chemicals (St. Louis, MO). [N-Methyl-CD$_3$]-morphine was purchased from Cerilliant (Round Rock, TX, cat. M-006).

3 Bacteria and Vectors

E. coli Plus KJ (Expression Technologies Inc.) is an *E. coli* B strain with a λ prophage carrying the T7 RNA polymerase gene under the lacUV5 promoter and a pACYC177-derived chloramphenicol-resistant plasmid expressing *E. coli* dnaK and dnaJ proteins. Genotype: F$^-$ *hsd*S *gal omp*T (λ cI857 ind-1 nin-5 Sam-7 *lac*UV5-T7 gene 1) *Lon*-.

E. coli DH5α (Clontech) was used as a second E. coli strain for the transformation of plasmid DNA. Genotype: F⁻ φ80lacZΔM15 Δ(□lacZYA-argF)U169 recA1 endA1 hsdR17(rk⁻, mk⁺) phoA supE44 thi-1 gyrA96 relA1 λ⁻

pET28a. The bacterial cloning and expression vector pET28a (Novagen), that was used in this thesis, is 5.6 kb large, carries a kanamycin resistance coding region and an N-terminal His•Tag®/thrombin/T7•Tag® configuration plus an optional C-terminal His•Tag sequence. The cloning/expression region of the coding strand is transcribed by T7 RNA polymerase.

4 Instruments

Electrophoresis: Minigel Twin G42 for SDS-PAGE and Agagel Mini for DNA agarosegel electrophoresis (Biometra)
PCR: 2720 Thermal Cycler (Applied Biosystems)
Radioactivity measurements: LS 6000 TA (Beckman coulter)
        automatic TLC-Linear Analyzer (Tracemaster 20, Berthold)
Centrifuges: 5810R Centrifuge (Eppendorf)
        5415D Centrifuge (Eppendorf)
        5415R Centrifuge (Eppendorf)
        Avanti J-25 (Beckman Coulter)
        Optima XL-100K Ultracentrifuge (Beckman Coulter)
Miscellaneous: pH meter SevenEasy (Mettler Toledo)
        IQ125 miniLab pH meter (IQ Scientific Instruments)
        UV/visible ultrospec 3000 Spectrophotometer (Pharmacia Biotech)
        Balance XS105 Dual Range and PB3002-S/Fact Delta Range (Mettler Toledo)
        Vortex mixer (Fisher Scientific)
        Water bath (Julabo)
        Freeze dryer (Virtis)
        Speed vac (H. Saur Laborbedarf)
        Rotavapor R-200 (Büchi)
        Tissue homogenizer (Ultra-Turrax T25 (IKA Labortechnik)
        Sonic Dismembrator Model 100 (Fisher Scientific)
        SPE Vacuum system Visiprep DL (Supelco)
        Air pump 1000 (Aquatic gardens)

5 Synthesis

5.1 Preparation of DOPAL from Epinephrine (according to DUNCAN 1975)

The synthesis of DOPAL from epinephrine was prepared as described by DUNCAN (1975) with some modifications. Briefly, 183 mg epinephrine were dissolved under stirring in 4 ml phosphoric acid (85-90%, Fluka cat. 79606) in a 10 ml round bottom flask while heated up to 120°C. After 120°C was reached the yellow reaction mixture was poured into 40 ml water that had previously been purged under a stream of nitrogen. The pH was adjusted to 3 with 5 N NaOH and the reaction was loaded onto a Strata X-C column (Phenomenex, 1g/12 ml, cat. 8B-S029-JDG) that had been equilibrated previously with 20 ml methanol and 20 ml 0.1 N HCl. The column was washed with 40 ml water and eluted with 10 ml methanol/acetonitrile 1:1 (v/v). Of each 1 ml-fraction 4 µl were loaded onto TLC and run in separation system 1 (II.10.1). DOPAL containing fractions were detected by spraying with 2,4-dinitrophenylhydrazine in 2 N HCl. The pooled fractions were dried under a stream of nitrogen, resuspended in ethylacetate and stored at -20°C (yield: 5.4%).

5.2 Synthesis of (R)-Norlaudanosoline*HBr

(R)-norlaudanosoline was prepared according to RICE & BROSSI (1980) from (R)-norreticuline. The reaction mixture consisted of 7 mg (R)-norreticuline*HCl in 150 µl 48% HBr and was dissolved while heating up in a silicone oil bath to 100°C within 30 min. After 2 h of incubation at 100°C the reaction was dried under a stream of nitrogen, resuspended in 50% methanol containing 0.2% acetic acid and subjected to preparative HPLC (II.10.3). Fractions containing (R)-Norlaudanosoline were pooled, evaporated with a rotary evaporator, shock-frozen in liquid nitrogen and lyophilized. (R)-Norlaudanosoline trifluoroacetate was obtained as a white powder in 77% yield. An aliquot was resuspended in water, loaded onto a TLC and separated using separation system 2 (II.10.1). The UV-absorbing band was scratched off and eluted with methanol. After filtration through a 0.2 µm Whatman PDF filter to remove silica residues the filtrate was evaporated under a stream of nitrogen and resuspended in 150 µl 24% HBr. After shock-freezing in liquid nitrogen and lyophilisation the purity was checked by HPLC (II.10.2) and LC-MS (II.10.4.A). (R)-Norlaudanosoline*HBr was obtained as a brown powder and immediately stored at -20°C.

5.3 Synthesis of [7D]-Salutaridinol and [7D]-*epi*-Salutaridinol

According to BARTON *et al.* (1965) and LENZ (1994) [7D]-salutaridinol and [7D]-*epi*-salutaridinol were synthesized from 80 mg salutaridine and 300 mg sodium borodeuteride in 4 ml methanol. The reaction mixture was stirred for 2 h at 0°C. After incubation for another 2 h at room temperature the reaction mixture was evaporated, resuspended in 8 ml water and extracted twice with 8 ml chloroform. Sodium sulphate was added to the pooled organic phase and it was let stand overnight. The organic phase was further filtrated and evaporated and two slightly different purification steps followed to separate [7D]-*epi*-salutaridinol from [7D]-salutaridinol.

a) Purification of [7D]-Salutaridinol

After resuspension in ethylacetate, [7D]-salutaridinol crystals were formed after digestion for 3 h at 50°C. These crystals were filtrated, recrystallized in methanol and subjected to TLC (II.10.1, separation system 3).

b) Purification of [7D]-*epi*-Salutaridinol

After resuspension in methanol, [7D]-*epi*-salutaridinol was purified by TLC in separation system 3 (II.10.1).

UV-absorbing bands corresponding to [7D]-salutaridinol or [7D]-*epi*-salutaridinol were eluted with methanol following filtration through a 0.2 µm Whatman PDF filter to remove silica residues. The filtrate was evaporated under a stream of nitrogen and resuspended in 0.01 N HCl. After shock-freezing in liquid nitrogen and lyophilisation, [7D]-salutaridinol hydrochloride and [7D]-*epi*-salutaridinol hydrochloride were resuspended in water, verified by HR-LC-MS (II.10.5) as shown in Fig. 3 and immediately stored at -20°C.

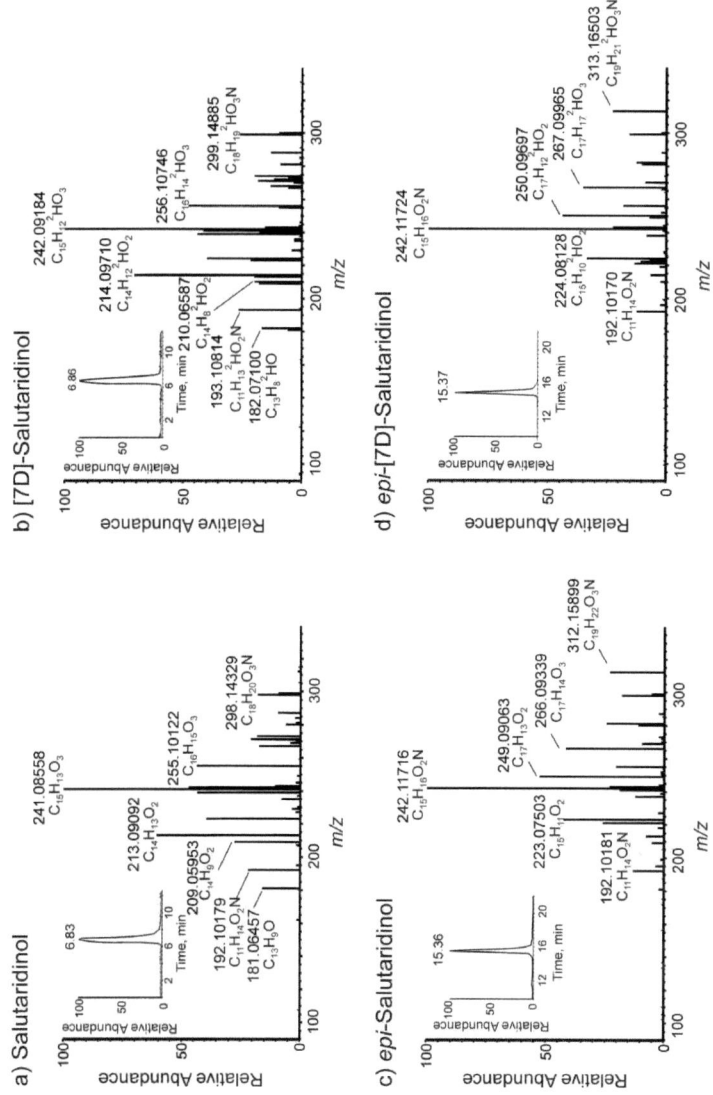

Fig. 3: HR-LC-MS analysis of chemically synthesized [7D]-salutaridinol and *epi*-[7D]-salutaridinol compared to unlabeled standards.

a) Full scan extracted chromatogram and MS/MS of salutaridinol.
b) Full scan extracted chromatogram and MS/MS of [7D]-salutaridinol.
c) Full scan extracted chromatogram and MS/MS of *epi*-salutaridinol.
d) Full scan extracted chromatogram and MS/MS of *epi*-[7D]-salutaridinol.

6 Application Experiments

6.1 Application to *Papaver setigerum* Seedlings

Sterilized 5 day old *Papaver setigerum* seedlings were incubated for 48 h with the potential precursor solution. The seedlings were extracted with 80% ethanol. The extracts were dried, redissolved in 20 µl 50% ethanol and separated by TLC in separation system 4 (II.10.1). Radioactive bands were detected by phosphorimaging with a Typhoon 9410 (Molecular Dynamics).

6.2 Injections into Mice and Collection of Urine

Up to five C57 black mice were put into a metabolic cage (kindly provided by Terry Sharp, Washington University, School of Medicine) with food and water *ad libitum* for one day to collect control urine. The potential precursor solution (50 nmol – 2000 nmol per 200 µl MilliQ water) was injected every 24 h for 4 days intraperitoneally using a 1 ml hypodermic syringe with a 26 gauge needle (1/2 inch) into each mouse. Mice were housed without food for 10 h post injection. Urine was collected every 24 h (1-2 ml per 5 mice) and immediately frozen at -20°C.

7 Preparation of Protein

7.1 Preparation of Crude Liver Protein

Fresh ice-cooled mouse liver tissue was weighed, cut into small pieces with a scalpel, disrupted with a 15 ml-Potter-Elvehjem homogenizer (Sartorius) in 3 vol 0.05 M Tris HCl pH 7.4 containing 0.15 M KCl and centrifuged for 20 min at 9000xg and 4°C. Protein concentration in supernatant and pelleted fraction was determined according to II.11.

7.2 Preparation of Microsomes from Liver

Microsomes were prepared according to KODAIRA & SPECTOR (1988) and FISINGER (1998). Supernatant from liver tissue was obtained as described in II.7.1 and transferred to an ultracentrifuge tube and centrifuged for 60 min at 43500 rpm or *ca*. 118000xg (Ti 90, Beckman coulter ultracentrifuge) and 4°C. Soluble protein was aliquoted, shock-frozen in liquid nitrogen and stored at -80°C. The microsomal pellet was carefully resuspended in 0.05 M Tris HCl pH 7.4 containing 0.15 M KCl with a 2 ml-Potter-Elvehjem homogenizer (B. Braun, Melsungen, Germany). Homogenized microsomes were aliquoted, shock-frozen in liquid nitrogen and stored at -80°C. Concentration of soluble and microsomal protein was determined according to II.11.

7.3 Heterologous Expression in *E.coli*

7.3.1 Cloning of Amine *N*-Methyltransferase cDNA in Vector pET28a and Transformation into *E. coli* PlusKJ

A 5 ml LB culture containing 30 µg/ml kanamycin (Luria-Bertani medium, SAMBROOK *et al.* 1989) of human amine *N*-methyltransferase (INMT) in vector pCR-BluntII-TOPO was prepared from a stab culture (Deutsches Ressourcenzentrum für Genomforschung, Germany) and incubated overnight at 37°C. After preparation of plasmid DNA according to the protocol of the QIAprep Spin Miniprep Kit (Qiagen) the sequence of the cDNA was confirmed by using following standard sequencing reaction with primers T7, M13rev, ORF_INMT_F1 and ORF_INMT_R792 (Tab. 2) and BigDye Termination Kit (Applied Biosystems):

Sequencing reaction		Sequencing reaction program (25 cycles)	
100-300 ng plasmid DNA	x µl	temperature	time, min
3.2 pmol/µl primer	1 µl	96°C	0:10
BigDye premix	3 µl	50°C	0:10
water	6-x µl	60°C	4:00
		4°C	∞

The analysis of the sequence was done at Washington University in St. Louis. Between all following steps 10 µl DNA sample was mixed with 4 µl 10x DNA loading buffer (50% (v/v) glycerol, 0.2 M EDTA, 0.05% (w/v) Orange G) and was checked by gel electrophoresis on a 1% (w/v) agarose gel. The size of DNA was estimated by loading a 1 kb DNA ladder (New England Biolabs). A standard PCR-reaction with *Pfu* DNA polymerase (Stratagene) was used to introduce *Xho*I and *Nhe*I restriction sites into pCR BluntII-TOPO at the 5'- and 3'-end of the INMT cDNA insert:

PCR-reaction		PCR-program (30 cycles)	
water	37 µl	temperature	time, min
10x *Pfu* DNA polymerase buffer	5 µl	94°C	3:00
1 ng plasmid DNA	1 µl	94°C	0:30
10 pmol/µl primer INMT_XhoI	2.5 µl	64°C	0:30
10 pmol/µl primer INMT_NheI	2.5 µl	72°C	2:00
10 nmol/µl dNTPs	1 µl	4°C	5:00
Pfu DNA polymerase	1 µl	4°C	∞

The PCR reaction was purified by phenol-chloroform-isoamylalchohol extraction (PCI extraction) following ethanol precipitation (EtOH precipitation). For PCI extraction, the DNA solution (vector pET28a or cDNA insert) was mixed 1:1 with phenol-chloroform-

Tab. 2: Primers that were used for cloning of amine *N*-methyltransferase cDNA into pET28a.

Primer	Sequence
T7	5'-TAATACGACTCACTATAGGG-3'
M13rev	5'-TTCACACAGGAAACAGCTATGACC-3'
ORF_INMT_F1	5'-ATGAAGGGTGGCTTCACTGGG-3'
ORF_INMT_R792	5'-TCAGGGCCCAGGCTTCTTGCGA-3'
INMT_XhoI	5'-AAACTCGAGTCAGGGCCCAGGCTTCTTGCGAG-3'
INMT_NheI	5'-GGGGCTAGCATGAAGGGTGGCTTCACTG-3'

isoamylalcohol (25:24:1), vortexed and centrifuged for 5 min at 13200 rpm and room temperature. The upper phase was mixed 1:1 with isoamylalchohol/chloroform (1:24), vortexed and centrifuged again. The upper phase was then mixed with 1/10 vol 3M sodium acetate pH 5.2 and 2.5 vol EtOH and placed for 5 min at -80°C. After centrifugation at 4°C and 13200 rpm for 20 min (table centrifuge 5415R, Eppendorf) the DNA pellet was washed with 200 µl 70% ethanol and centrifuged again for 5 min. The DNA pellet was dried and then incubated overnight at 37°C for sequential restriction digest first with *XhoI* (Invitrogen) and then *NheI* (New England Biolabs). The incubation conditions were according to the protocol described by the manufacturer. The final cut vector pET28a was purified by PCI extraction and EtOH precipitation; the cut cDNA insert was purified by gel extraction with QIAprep MiniElute Gel Extraction Kit (Qiagen) and EtOH precipitation. Following ligation reaction with T4 DNA ligase (Promega) was incubated overnight at 16°C:

200ng/ µl cut cDNA insert	7 µl
200ng/ µl cut vector pET28a	1 µl
10x T4 DNA ligase buffer	1.4 µl
T4 DNA ligase	0.5 µl

The ligation reaction was added to 100 µl competent cells *E. coli* DH5α (prepared by Dr. Taiji Nomura) and placed on ice for 30 min. The reaction mixture was incubated for 45 sec at 42°C, placed on ice for 2 min and added to 900 µl LB media. After incubation for 1 h at 37°C the reaction was centrifuged at 10000 rpm for 30 sec. Then, 800 µl of the supernatant was removed and the cells were resuspended with 200 µl LB-medium. The resuspended cells were divided into two portions (20 µl and 180 µl) and plated on LB agar plates containing 50 µg/ml kanamycin. After incubation overnight at 37°C, colonies were transferred with a pipette tip into following PCR-reaction mixture (colony-PCR) containing *Taq* DNA polymerase (Invitrogen):

PCR-reaction		PCR-program (25 cycles)	
water	21.2 µl	temperature	time, min
10x *Taq* DNA polymerase buffer	3 µl	94°C	2:00
50 mM MgCl$_2$	0.9 µl	94°C	0:30
10 pmol/µl INMT_XhoI	2 µl	55°C	0:30
10 pmol/µl INMT_NheI	2 µl	72°C	1:00
10 nmol/µl dNTPs	0.6 µl	4°C	4:00
Taq DNA polymerase	0.3 µl	4°C	∞

After mini-preparation and sequencing of the pET28a vector carrying the NMT cDNA insert a transformation into the expression strain *E. coli* Plus KJ (Expression Technologies Inc) followed. Glycerol stocks of *E. coli* DH5α pET28a_NMT and *E. coli* Plus KJ pET28a_NMT were prepared by mixing 1 ml fresh overnight bacteria culture with 50% glycerol and stored at -80°C.

7.3.2 Isolation of Amine *N*-Methyltransferase His-tagged Protein from *E. coli*

One-liter-culture of *E. coli* Plus KJ pET28a_NMT in LB containing 50 µg/ml kanamycin was grown for 5 h until an OD$_{600}$ of 0.6-0.8. Protein expression was induced by adding IPTG to a final concentration of 1 mM. After incubation for 16 h at 28°C the culture was centrifuged at 8000xg and 4°C for 10 min. The pellet was resuspended in 20 ml Histag-buffer (50 mM Tris/HCl pH 7.5 containing 500 mM NaCl, 10% (v/v) glycerol, 2.5 mM imidazol and freshly added 5 mM β-mercaptoethanol). After adding 1 mg/ml lysozyme to the resuspended cells an incubation for 40 min at 4°C followed. The cells were disrupted by sonication 5x2 min (continuous, step 5, cool in between 5 min on ice). During centrifugation at 20000xg for 20 min at 4°C Talon resin (Clontech) was prepared by spinning down 2 ml of the resin at 700xg for 2 min (Eppendorf). The resin was washed twice with 10 ml His-tag buffer, resuspended with crude protein extract and incubated for 40 min at 4°C. After centrifugation at 700xg and 4°C for 5 min, unbound protein was discarded and the resin was washed twice for 10 min with His-tag buffer. The resin was resuspended in 2 ml His-tag buffer and transferred to a disposable gravity column (Clontech) and left to settle for 30 min. Bound protein was washed with 5 ml His-tag buffer and eluted with 5 ml His-tag buffer containing 150 mM imidazol. Eluted His-tagged protein was desalted with storage buffer (20 mM Tris/HCl pH 7.5, 150 mM NaCl, 10% (v/v) glycerol, 5 mM β-mercaptoethanol) on a PD10 column (GE Healthcare) according to the manufacture's protocol and stored at -20°C. The concentration of the protein was determined as described in II.11. The purity of the protein sample was analyzed on a SDS polyacrylamide gel prepared as follows:

running gel		stacking gel	
1.5 M Tris/HCl pH 8.8	2.5 ml	0.5 M Tris/HCl pH 6.8	1.25 ml
40% Acrylamide	3 ml	40% Acrylamide	0.5 ml
20% SDS	50 µl	20% SDS	25 µl
10% APS	50 µl	10% APS	50 µl
TEMED	15 µl	TEMED	10 µl
Glycerol	0.5 ml	water	3.165 ml
water	3.935 ml		

For SDS-PAGE in SDS running buffer (25 mM Tris, 2M glycine, 1% (w/v) SDS) the protein sample was mixed with 5x SDS sample buffer (150 mM Tris/HCl pH 6.8, 10% (w/v) SDS, 25% (w/v) sucrose, 0.01% (w/v) bromphenol blue, 25% (v/v) β-mercaptoethanol) and denaturated at 100°C for 5min. A protein marker as a standard (Bio-Rad) to estimate the protein size was also loaded. The SDS-polyacrylamide gel was stained for 30 min in Coomassie staining solution (0.25% (w/v) Coomassie Brilliant Blue G250, 5% (v/v) methanol, 7.5% (v/v) acetic acid) and destained in 7% acetic acid (v/v).

7.3.3 Thrombin Digestion of His-tagged Protein

For the removal of the His-tag from the protein a dialysis membrane (VISKING Typ 1 7/8 ss, Roth, cat. 5358.1) with a molecular weight cut-off of 14000 kDa was first washed in 1 l MilliQ water for 2x30 min. All steps were conducted at 4°C. Digestion was started by adding 0.5 µl of thrombin solution (Sigma Aldrich, 1 unit/ml) to 1 mg of His-tagged protein (1/6400 the amount of NMT by weight, JEZ & CAHOON 2004). The reaction mixture was carefully transferred into the dialysis membrane, which had to be kept moistened. After dialysis for 2x12 h in 1 l storage buffer (II.7.3.2) the dialyzed sample was transferred into an Eppendorf tube and a resin mixture consisting of 100 µl Talon (Clontech) and 80 µl Benzamidine Sepharose (GE Healthcare) was first washed with storage buffer and then added to the dialyzed sample to remove cleaved His-tag, uncleaved protein and thrombin. After incubation for 20 min on ice the mixture was centrifuged at 700xg for 5 min. Protein was determined as described in II.11 and stored until further use at -20°C.

8 Standard Test for Mammalian Liver Enzymes

According to KODAIRA & SPECTOR (1988) enzyme assays with microsomal and cytosolic mammalian liver enzyme were conducted in 0.05 M potassium phosphate buffer pH 7.4 in a total volume of 1 ml containing 10 µM substrate (10 nmol), 5.5 mM glucose-6-phosphate (5.5 µmol), 1 unit glucose-6-phosphate dehydrogenase, 1 mM NADP$^+$ (1 µmol), 5 mM MgCl$_2$ (5 µmol), 1 mM NADH (1 µmol) and 1-3 mg microsomal or cytosolic liver protein. The reaction was terminated after an incubation of 2 h at 37°C with 100 µl 20% TCA.

9 Solid Phase Extraction of Alkaloids

9.1 Strata X-C (Phenomenex)-TCA method

The sample was homogenized in 6 ml 5% (w/v) TCA using a tissue homogenizer (Ultra-Turrax T25, IKA Labortechnik) following a centrifugation step of 15 min at 3000xg and 4°C. The supernatant was transferred into a clean tube and the pellet was homogenized with 3 ml 5% (w/v) TCA and centrifuged for a second time. The supernatants from both steps were combined and loaded onto a Strata X-C (Phenomenex, 500 mg, 6 ml, cat. 8B-S029-HCH) that had been preconditioned with 1 vol methanol, 1 vol 0.1 N HCl and 2 vol water. The cartridge was rinsed with 10 ml 0.1 N HCl, 10 ml 0.1 N HCl/5% methanol, 15 ml 60% methanol/ acetonitrile (1:1), 15 ml methanol/acetonitrile (1:1) and eluted with 6 ml 0.625% ammonium hydroxide in methanol. The elution fraction was evaporated under a stream of nitrogen, resuspended in 50 µl methanol and subjected to TLC in separation system 5 (II.10.1), LC-MS (II.10.4.A) or scintillation counting (II.12).

9.2 Strata X-C (Phenomenex)-HClO$_4$ method

After homogenizing the sample in 7.5 ml of a 0.2 N HCLO$_4$ mixture containing 0.4 mM sodium sulfite and 0.4 mM EDTA using a tissue homogenizer a centrifugation step followed for 10 min at 3000xg and 4°C. The pH of the supernatant was adjusted to 7 with 0.2 N ice-cold KOH. After a second centrifugation step 2 N HCl was added to the supernatant to a final concentration of 0.1 N HCl. The sample was loaded onto a Strata X-C (Phenomenex, 500 mg, 6 ml, cat. 8B-S029-HCH) that had been preconditioned with 20 ml methanol and 20 ml 0.1 N HCl. The cartridge was rinsed with 10 ml 0.1 N HCl, 10 ml 0.1 N HCl/5% methanol, 15 ml 60% methanol/ acetonitrile (1:1), 15 ml methanol/acetonitrile (1:1) and eluted with 6 ml 2% ammonium hydroxide in methanol. The elution fraction was evaporated under a stream of nitrogen, resuspended in 50 µl methanol and subjected to TLC in separation system 5 (II.10.1), LC-MS (II.10.4.A) or scintillation counting (II.12).

9.3 Bond Elut Certify (Varian)

Work-up of urine was conducted either in St. Louis or in the group of Prof. Michael Spiteller, Dortmund, Germany, according to HOFMANN *et al* (1999) with slight modifications. After hydrolysis of an aliquot of urine (maximum 5 ml) with 37% HCl (final concentration 2 N) for 40 min at 110°C, the mixture was cooled down and pH was adjusted to 7-8 with 10 N KOH. The sample was loaded onto a Bond Elut Certify cartridge (Varian, 130 mg, 3 ml, cat. 12102051) that had been preconditioned with 2 ml methanol and 2 ml water. The cartridge was rinsed with 2 ml water, 1 ml acetate buffer pH 4.0, 2x2 ml methanol and, after 2 min vacuum, eluted with 2 ml dichloromethane/isopropanol/ammonium hydroxide (8:2:0.2). The elution fraction was evaporated under a stream of nitrogen, resuspended in 100 µl water/methanol (8:2) containing 0.1% formic acid and subjected to HR-LC-MS (II.10.5).

9.4 Sep Pak Plus C18 (Waters)-basic conditions

According to a suggestion from Dr. Baichen Zhang, Donald Danforth Plant Science Center, St. Louis, USA, a fast SPE-method was developed to extract alkaloids from a biological sample under basic conditions. Brain tissue was hydrolyzed at 110°C for 40 min in 1.8 ml 0.2 N HCl (1.5 ml water + 0.3 ml 37% HCl). After adjusting the pH of the sample to 8-8.5 with 1 N KOH it was loaded onto a Sep Pak Plus C18 cartridge that had been preconditioned with 2x 1ml methanol, 2x 1ml water, 2x 1ml 0.5% ammonium hydroxide in water. The cartridge was rinsed 2x with 1ml methanol and after 1 min vacuum eluted with 5 ml 0.5% ammonium hydroxide in methanol. The sample was evaporated under a stream of nitrogen and subjected to LC-MS (II.10.4).

10 Chromatographic methods

10.1 Thin Layer Chromatography (TLC)

Alkaloids were separated by TLC on Polygram silica G/UV254 plates (Macherey-Nagel, layer: 0.2 mm silica gel with fluorescent indicator UV254). UV-absorbing bands were visualized at 254 nm with a Min UVIS (Bachofer). Following solvent systems (all v/v) were used:

1) ethylacetate/n-hexane, 2:1,
2) butanol/acetic acid/water, 4:1:1
3) chloroform/acetone/diehtylamine 5:4:1
4) toluene/ethyl acetate/diethylamine, 7:2:1
5) methanol/ammonium hydroxide, 100:1.5
6) chloroform/ethylacetate/ethanol/ammonium hydroxide, 6:1:2:1
7) toluene/acetone/ethanol/ammonium hydroxide, 45:45:7:3
8) chloroform/methanol/ammonium hydroxide, 90:9:1

10.2 Analytical High Performance Liquid Chromatography

The HPLC system consisted of L-7100 HPLC pump and L-7200 autosampler (Merck Hitachi). UV-data was obtained with an L-7450 diode array detector (Merck Hitachi). Separation of (10 µl) samples was achieved by using a LiChroCART 250-4 HPLC column (Merck, 5 µm, LiChrosphor 60 RP select B) combined with a LiChroCART 4-4 HPLC cartridge (Merck, 5 µm, LiChrosphor 60

RP select B). The mobile phase total flow was set to 0.8 ml/min with binary gradient elution, using 0.1% TFA as solvent A and acetonitrile as solvent B. The gradient started with 5% B, was increased to 50% B over 14 min and then 100% B over 2 min. Elution was continued for 2 min at 100% B followed by a 8-min equilibration with the starting condition. Fixed wavelengths were set to 230, 245, 280 and 320 nm.

10.3 Preparative High Performance Liquid Chromatography

The HPLC system consisted of L-7100 HPLC pump and L-7200 autosampler (Merck Hitachi). UV-data was obtained with an L-7450 diode array detector (Merck Hitachi). Separation of (400 µl) samples was achieved by using a Hibar Pre-Packed column RT250-25 (Merck, 7 µm, LiChrosorb RP-18). The mobile phase total flow was set to 8 ml/min with binary gradient elution, using solvent A (0.1% TFA) and B (acetonitrile). The gradient started with 10% B and was increased to 30% B over 60 min followed by extensive washing with 100% B for 30 min and a 30-min equilibration with the starting condition. Fixed wavelengths were set to 230, 245, 280 and 320 nm.

10.4 Liquid Chromatography-Mass Spectrometry (LC-MS)

The system consisted of a CTC Pal autosampler (LEAP Technologies), a Shimadzu LC-20AD liquid chromatograph and a 4000 QTRAP mass spectrometer (Applied Biosystems). Compound-dependent parameters are described in Tab. 3. Three HPLC separation systems were used dependent on the analytical problem and analyte. The following TIS source parameters were used: CUR 30, CAD high, IS 5000, TEM 500, EP 10, dwell time 50 ms.

Tab. 3: Compound-dependent parameters for the LC-MS/MS method.

Analyte	Collision energy (V)	Declustering potential (V)	Quantifier MRM transition	Qualifier MRM transition	CXP
DOPA-pyruvate	-15	-30	195 → 123	195 → 151	-17
DOPAL	-25	-70	151 → 123	151 → 122	-6
DOPET	-30	-62	153 → 123	153 → 122	-7
norlaudanosoline	30	70	288 → 164	288 → 123	17
laudanosoline	26	40	302 → 178	302 → 123	17
4'-O-methylnorlaudanosoline	33	35	302 → 164	302 → 285	17
4'-O-methyllaudanosoline	30	40	316 → 178	316 → 123	17
6-O-methylnorlaudanosoline	30	40	302 → 178	302 → 285	17
6-O-methyllaudanosoline	30	40	316 → 192	316 → 123	17
norreticuline	30	45	316 → 178	316 → 299	17
reticuline	35	45	330 → 299	330 → 192	17
corytuberine	40	50	328 → 265	328 → 282	17
pallidine	40	50	328 → 211	328 → 237	17
salutaridine	40	50	328 → 211	328 → 237	17
isoboldine	40	50	328 → 265	328 → 237	17
thebaine	35	40	312 → 251	312 → 281	17
oripavine	35	40	298 → 218	298 → 249	17
codeine	40	47	300 → 215	300 → 225	17
morphine	45	45	286 → 201	286 → 165	17

A) **Analysis of DOPAL, DOPET, THBIQ and Morphine Alkaloids:**

Separation of (10 µl) samples was achieved by using an Eclipse XDB-C18 HPLC column (Agilent, 2.1x150 mm, 3.5 µm) combined with a microfilter unit (Sigma, 1/16 OD tubing, Cat# 502693). The mobile phase total flow was set to 0.2 ml/min or 0.3 ml/min with binary gradient elution, using solvent A (0.1% formic acid) and B (Acetonitrile). Following gradients were used depending on the analyte:

analyte: **DOPAL, DOPET**		analyte: **THBIQ**		analyte: **morphine alkaloids**	
time, min	solvent B,%	time, min	solvent B,%	time, min	solvent B,%
2	20	2	5	2	10
10	100	6	100	8	100
12	100	10	100	10	100
13	20	12	5	12	10
15	20	15	5	15	10

B) Chiral Separation of THBIQ Alkaloids:

Separation of (10 µl) samples was achieved by using a Chiral-CBH HPLC column (Chromtech, 100x4.0 mm) combined with a Chiral-CBH guard cartridge (Chromtech, 10 x 4 mm). The mobile phase total flow was set to 0.9 ml/min with isocratic elution using 5% acetonitrile, 10 mM ammonium acetate pH 5.5. Between each run the column had to be generated with 5% acetonitrile, 10 mM ammonium acetate pH 5.5 containing 50 µM EDTA (with diverter set to waste).

C) Analysis of the Phenol-coupled Products:

Separation of (10 µl) samples was achieved by using a Luna C18 octadecylsilane HPLC column (Phenomenex, 5 µm, 150 mm × 2 mm) combined with a C18 guard column (Phenomenex, 4x2 mm). The mobile phase total flow was set to 0.5 ml/min with binary gradient elution, using solvent A (5% methanol, 5% acetonitrile, 10 mM ammonium bicarbonate, 45 mM ammonium hydroxide) and B (90% acetonitrile, 10 mM ammoniumbicarbonate, 15 mM ammoniumhydroxide) (all v/v). The gradient started with 100% A for 2 min and was increased to 100% B over 10 min. Elution was continued for 2 min at 100% B followed by a 5-min equilibration with the starting condition.

10.5   Liquid Chromatography-High Resolution Mass Spectrometry (HR-LC-MS)

Analysis was done in the group of Dr. Michael Spiteller by Dr. Marc Lamshöft at the Institute for Environmental Research in Dortmund, Germany. The APCI-FT-MS spectra were obtained using an LTQ-Orbitrap Spectrometer (Thermo Fisher, USA). The spectrometer was operated in positive mode (1 spectrum s^{-1}; mass range: 50-1000) with a nominal mass resolving power of 60000 at m/z 400 at a scan rate of 1 Hz using automatic gain control to provide high-accuracy mass measurements ($\leq$ 2 ppm deviation). For the determination of elemental composition the internal calibration standard bis-(2-ethylhexyl)-phthalate (m/z 391.28428) was used. The spectrometer was equipped with a Surveyor HPLC system (Thermo Scientific, USA) consisting of LC-Pump, UV detector (λ = 254 nm) and autosampler (injection volume 10 µl). Separation of samples was achieved by using a Synergi Fusion RP HPLC column (Phenomenex, 4 µm, 150 x 3 mm) combined with a Synergi Fusion RP guard column (Phenomenex, 4 x 3 mm). The mobile phase total flow was set to 0.5 ml/min with binary gradient elution, using solvents A (0.1% formic acid, 10mM ammonium acetate) and B (0.1% formic acid in acetonitrile) (all v/v). The gradient started with 5% B for 4 min and was increased to 30% B over 20 min. Elution was continued for 10 min at 100% B followed by a 7-min equilibration with the starting condition. Identification of metabolites was initiated by Dr. M. Lamshöft and later confirmed and expanded by myself. To analyze the data qualitatively and quantitatively raw files were transferred *via* FileZilla Cient connecting the server of the Institute of Environmental Research with a desktop client of the Donald Danforth Plant Science Center.

11 Protein Assay

Concentration of protein was determined according to BRADFORD (1976) by mixing 20 µl of sample with 1 ml 1:5 diluted Bradford reagent (Bio-Rad Laboratories). Bovine serum albumin was used as external standard. After incubation for 5 min at room temperature the extinction at 595 nm was measured with a UV/visible Spectrophotometer (Ultrospec 3000, Pharmacia Biotech).

12 Radioactivity Measurements

The solution containing ^{14}C- or ^{3}H-label was mixed with 5 ml scintillation cocktail (Bio-Safe II, Research Products International Corp.) and quantitatively analyzed with a scintillation counter (LS 6000 TA, Beckman Coulter). TLCs of ^{14}C- or ^{3}H-labeled compounds were qualitatively and quantitatively analyzed with an automatic TLC-Linear Analyzer (Tracemaster 20, Berthold) and a Phosporimager (Typhoon 9410, Molecular Dynamics).

III Results

Part A: Morphine analytics

Detection of morphine in animal organs and body fluids has been attempted for more than 30 years (Tab. 1). The analytical methods that were chosen such as RIA, ELISA and EC-HPLC take advantage of great sensitivity but lack the ability to unequivocally confirm the presence of alkaloids in a biological sample. Mass Spectrometry is the analytical method of choice to resolve this problem because of its capability to selectively verify a molecule and to discriminate it against matrix, background and other chemical species. The great specificity of MS due to a mass-based detection also provides additional structural information consistent with high sensitivity. Two main extraction procedures have been described for successful isolation of morphine from biological samples: liquid-liquid and solid-phase extraction. However, since recoveries for liquid-liquid extraction are usually not high enough, solid-phase extraction (SPE) alone or a combination of both techniques are used for sample preparation. One strategy for SPE is reversed-phase chromatography that uses hydrophobic interactions of the analyte with a non-polar stationary phase. Another strategy is cation exchange chromatography based on the analyte's characteristic to build positively charged ions under basic conditions. A combination of both chromatographic principles is unified in the SPE cartridge Strata X-C (Phenomenex). Described in this thesis are studies with radio-labeled [N-methyl-^{3}H]-morphine and heavy-isotope labeled [N-methyl-CD_3]-morphine that show new insight into morphine analytics including examination of morphine's stability in different chemical environments, evaluation of the SPE method in the example of Strata X-C SPE cartridges and challenges of MS detection as determined with a 4000 QTRAP MS instrument (Applied Biosystems).

1 Stability of morphine

For our analysis we obtained a standard solution of [N-methyl-^{3}H]-morphine from American Radiolabeled Chemicals (80 Ci/mmol, 1 mCi/ml cat. ART 0659, stored at -80°C). From that standard solution a stock solution of 5 µCi [N-methyl-^{3}H]-morphine in 10 ml 75% ethanol was prepared and stored at 4°C. For each experiment 0.1–0.2 µCi (120000-240000 cpm, 1.25-2.5 pmol, 356-712 pg) [N-methyl-^{3}H]-morphine were used. We also obtained a standard solution of [N-methyl-CD_3]-morphine in methanol (Cerilliant, 1 mg/ml, cat. M-006, stored at 4°C). From that standard solution a stock solution of 10 µg/ 10 ml [N-methyl-CD_3]-morphine in methanol was prepared (stored at 4°C) of which 17-35 pmol (5-10 ng) were used for each experiment.

For most of the extraction procedures it is required that morphine is concentrated or reconstituted in another solvent by evaporation. First, it was analyzed whether morphine can be recovered equally from glass tubes or polypropylene tubes. For that, [N-methyl-CD_3]-morphine in methanol was provided in glass vials and evaporated under a stream of nitrogen or with a rotary evaporator. As determined by LC-MS (II.10.4.A) only 44% morphine was recovered from a glass vial after evaporation and reconstitution in methanol. An overnight silanization of glass vials with Sigmacote (Sigma, cat. SL2-100ml) improved recovery and 74% of [N-methyl-CD_3]-morphine could be reconstituted with methanol from the silanized glass vial. From a polypropylene tube, 90% [N-methyl-CD_3]-morphine could be instead recovered after evaporation and reconstitution with methanol as determined by LC-MS (II.10.4.A). Based on these results obtained with heavy-isotope labeled morphine, polypropylene test vials were used for the following reconstitution experiments with radio-labeled morphine. For that, 0.1 µCi (120000 cpm) [N-methyl-^3H]-morphine were diluted in methanol of which an aliquot of 0.006 µCi (7000 cpm) was evaporated. After reconstitution with methanol 100% [N-methyl-^3H]-morphine were recovered as determined by scintillation counting (II.12).

To evaluate both detection systems, scintillation counting as well as LC-MS, 0.1 µCi [N-methyl-^3H]-morphine and 60 ng [N-methyl-CD_3]-morphine were mixed together. After evaporation and reconstitution with methanol 93% [N-methyl-^3H]-morphine and 96% [N-methyl-CD_3]-morphine were recovered as determined by scintillation counting (II.12) and LC-MS (II.10.4.A), respectively. Additionally, it was shown by separating an aliquot of reconstituted [N-methyl-^3H]-morphine by TLC in the basic solvent system 5 containing methanol and ammonium hydroxide (II.10.1), that the reconstituted compound was not degraded and that after evaporation with air or twice with nitrogen 93% [N-methyl-^3H]-morphine were recovered.

In these previous experiments methanol was used for reconstitution. The evaluation of several other possible solvents for the reconstitution of [N-methyl-^3H]-morphine from a polypropylene tube was conducted by charging the test tubes with 0.01 µCi (12000 cpm) [N-methyl-^3H]-morphine. After evaporation the samples were left for seven days at room temperature. Radio-labeled morphine was then reconstituted with ten different solvents. A quantitation by scintillation counting (II.12) revealed that among the ten different conditions tested methanol, 0.1 N HCl as well as methanol or ethylacetate containing 0.1 N HCl were

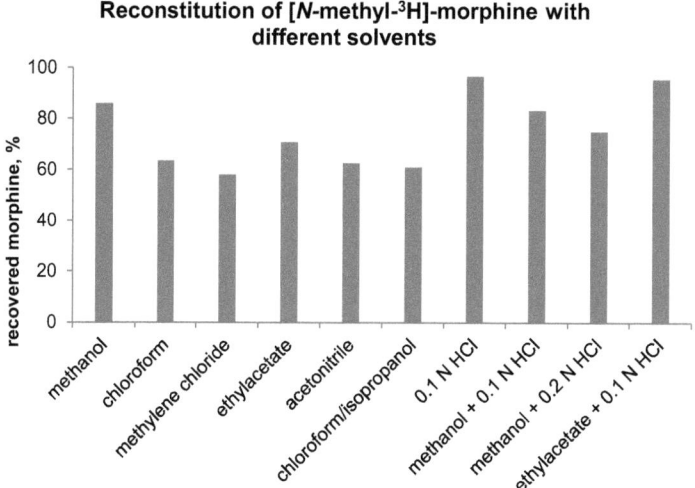

Fig. 4: Reconstitution of [*N*-methyl-^{3}H]-morphine from polypropylene tubes with 10 different solvents. Quenching of solvents especially chloroform (95%), methylene chloride (58%) and chloroform/isopropanol 3:1 (91%) was accounted for in these calculations by adding again 0.01 µCi (12000 cpm) [*N*-methyl-^{3}H]-morphine to the resuspended samples. The difference between both radioactivity measurements was set 100%.

found to be the best solvents for the reconstitution of morphine from polypropylene tubes (Fig. 4). For detection by MS, however, only a methanol diluted sample can be directly injected without taking any further precautions.

With the goal to use Strata X-C SPE for the extraction of morphine from animal tissue, the stability of morphine under basic conditions that are required for the elution from the column needed to be further analyzed. For that, the two different concentrations of 0.625% and 2% ammonium hydroxide in methanol were used. After 0.2 µCi (240000 cpm) [*N*-methyl-^{3}H]-morphine was evaporated and resuspended in methanol containing either 0.625% or 2% ammonium hydroxide, the solution was left at room temperature for 7 days. An aliquot of 0.01 µCi (12000 cpm) was loaded onto TLC every 24 h, separated in the basic solvent system 5 containing methanol and ammonium hydroxide (II.10.1) and quantitatively as well as qualitatively analyzed with a TLC-Linear Analyzer (II.12). As it is depicted in Fig. 5, [*N*-methyl-^{3}H]-morphine was not degraded after evaporation and reconstitution and morphine was found to be stable in 0.625% or 2% ammonium hydroxide for 7 days.

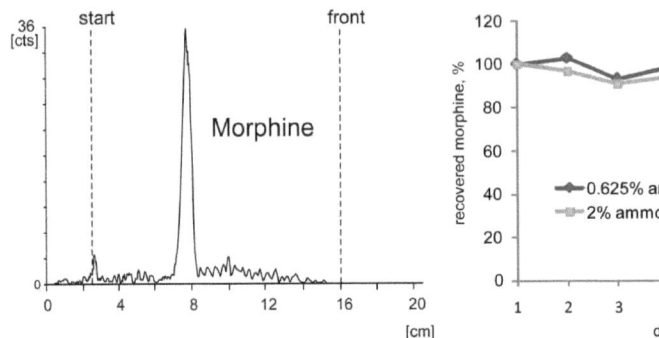

Fig. 5: Stability of [*N*-methyl-^{3}H]-morphine in basic conditions.

a) Radio-TLC in separation system 5 (II.10.1) of 0.01 µCi or 12000 cpm [*N*-methyl-^{3}H]-morphine that was evaporated and reconstituted in 2% ammonium hydroxide/methanol.
b) Stability of [*N*-methyl-^{3}H]-morphine kept at room temperature for 7 days in diluted ammonia (0.625% and 2%) corrected for the impurity that is shown in Fig. 5a).

2 Solid phase extraction with Strata X-C

Strata X-C (Phenomenex, 500 mg, 6 ml, cat. 8B-S029-HCH) and the protocol described in II.9.1 utilizing TCA as an acidifying and protein precipitating agent were chosen to investigate the SPE method's potential of isolating morphine from biological background (animal tissue). For all following experiments 0.1 µCi (120000 cpm, 1.25 pmol, 356 pg) [*N*-methyl-^{3}H]-morphine were used unless otherwise stated. First, analysis was performed to determine whether the non-activated material of the column retains any morphine or if ideally all spiked morphine can be found in the effluent. For that, Strata X-C was washed with methanol containing 0.625% ammonium hydroxide. Radio-labeled morphine diluted in the 0.625% ammonium hydroxide was loaded onto the column and expected not to bind to the stationary phase. The effluent was immediately collected and analyzed by scintillation counting (II.12) and, indeed, [*N*-methyl-^{3}H]-morphine was detected in the effluent with a recovery of 100%.

Next, analysis was performed to see whether activated Strata X-C is able to retain morphine until its elution under basic conditions. The standard protocol for activation of Strata X-C as described in II.9.1 was conducted. After radio-labeled morphine was loaded onto the column several washing steps followed containing 0.1 N HCl, 0.1 N HCl/5% methanol, 60% methanol/acetonitrile (1:1) and methanol/acetonitrile (1:1). It was observed that activated Strata X-C retained bound [*N*-methyl-

^{3}H]-morphine for all washing steps to which the column was exposed and 100% were recovered in the elution fraction as determined by scintillation counting (II.12).

In addition, using a tissue homogenizer (Ultra-Turrax T25, IKA Labortechnik), hydrolysis of the sample at 110°C for 40 min as well as changing the loading conditions from 5% TCA to 0.1 N or 1 N HCl did not lead to a decrease in recovery of [N-methyl-^{3}H]-morphine. It was further observed that [N-methyl-^{3}H]-morphine bound to Strata X-C for 5 days was not found to be degraded after elution with 2% ammonium hydroxide in methanol as determined by TLC in the basic separation system 5 containing methanol and ammoniumhydroxide (II.10.1) and scintillation counting (II.12).

Two different concentrations of ammonia in methanol (0.625% and 2%) were examined for the ability to elute radio-labeled morphine from Strata X-C columns. As shown in Fig. 6 methanol containing 0.625% or 2% ammonium hydroxide is sufficient to elute 0.01 µCi [N-methyl-^{3}H]-morphine (12000cpm, 0.125 pmol, 35.6 pg) from the column immediately after binding. A recovery of 83% and 100% was determined by scintillation counting (II.12) for the elution with 0.625% and 2% ammonium hydroxide in methanol, respectively.

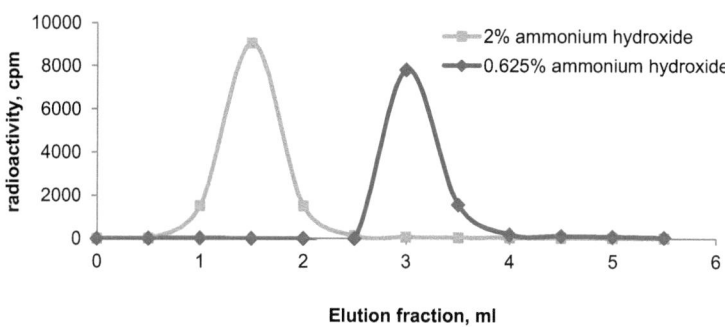

Fig. 6: Elution of [N-methyl-^{3}H]-morphine from Strata X-C with two different concentrations of diluted ammonia in methanol (0.625% and 2%).

Further testing was performed using two other Strata X-C columns available from Phenomenex with different dimensions: "Mini-Strata" (30 mg, 1 ml, cat. 8B-S029-TAK) and "Giga-Strata" (1g, 12 ml, cat. 8B-S029-JDG). With both columns a recovery of 117%-118% [N-methyl-^{3}H]-morphine was observed after SPE as described in II.9.1 utilizing TCA as the protein preicipitating agent.

With the goal to purify morphine from animal tissue or body fluids it was also evaluated whether biological background such as brain tissue interferes with the recovery of morphine from Strata X-C. For that, 1 µCi [N-methyl-^{3}H]-morphine (1200000 cpm, 12.5 pmol, 3560 pg), were directly injected with a Hamilton syringe into freshly thawed mouse brain tissue (C57 black mouse, ca. 450 mg FW). After immediate SPE work-up of the tissue as described in II.9.1 utilizing TCA as the protein precipitating agent, a recovery of 98% completely intact [N-methyl-^{3}H]-morphine was observed. This recovery was determined by scintillation counting (II.12) and TLC in the basic solvent system 5 containing methanol and ammonium hydroxide (II.10.1).

Finally, the results obtained with biological background were confirmed with a second SPE method for Strata X-C, which is described in II.9.2 and that differs only in the agent perchloric acid that is used instead of TCA for precipitating protein. Again, a 100% recovery of 1 µCi (1200000 cpm, 12.5 pmol, 3560 pg), that was directly injected prior SPE with a Hamilton syringe into two freshly thawed mouse brains (C57 black mice, ca. 888 mg total FW), was achieved as determined by scintillation counting (II.12).

3 Ion suppression

Since experiments with Strata X-C showed a 100% recovery of the isotopic tracer [N-methyl-^{3}H]-morphine, the standard SPE method (II.9.1) was applied to aqueous solutions that were additionally spiked with 17.4 pmol [N-methyl-CD_3]-morphine (5 ng) to analyze the recovery by LC-MS detection. Whereas 97%-104% of the radio-labeled morphine was recovered in these samples as determined by scintillation counting (II.12), only an average of 4% heavy-isotope labeled morphine was detected by MS (Tab. 4). Addition of mouse or rat brain tissue to the SPE procedure (II.9.1) decreased the recovery of [N-methyl-CD_3]-morphine even more to 0.7% as determined by LC-MS (Tab. 4). The yield of MS signal could also not be improved when the second SPE method (II.9.2) was used that differed in the protein precipitating agent (Tab. 4). Moreover, if unlabeled morphine was spiked into an elution fraction obtained after SPE, the same loss of MS signal as for heavy-isotope labeled morphine was observed for the unlabeled analyte suggesting that severe ion suppression of the analyte is the reason for these poor recoveries of less than 7%. To improve the recoveries, LC-MS conditions were subjected to change with the help of Dr. Baichen Zhang, Donald Danforth Plant Science Center, St. Louis, MO. The concentration of organic solvent in solvent A was decreased from 90% methanol to 10% and the pH was adjusted to pH 9 with ammonium hydroxide. The new solvent A consisted of 10% methanol containing 10 mM ammonium acetate and 10 mM ammonium hydroxide and was found to be beneficial for the separation of morphine.

Tab. 4: Recoveries of [N-methyl-CD$_3$]-morphine (MoCD$_3$) obtained by LC-MS after SPE with Strata X-C (before and after LC-method development).

Recovery of MoCD$_3$, without biological background	Recovery of MoCD$_3$, with biological background	Biological background	SPE method
before LC-method development:			
1%	0.9%	4x mouse brains (1.7 g FW)	II.9.1 (TCA)
7%	0.5%	2x rat brains (3.5 g FW)	II.9.1 (TCA)
5%	0.8%	2x rat brains (3.2 g FW)	II.9.2 (HClO$_4$)
(average: 4%)	(average: 0.7%)		
after LC-method development:			
38%	21%	1x mouse brain (0.3 g FW)	II.9.2 (HClO$_4$)
38%	24%	1x rat brain (1.6 g FW)	II.9.2 (HClO$_4$)
not determined	20%	2x rat brains (3.3 g FW)	II.9.2 (HClO$_4$)
not determined	41%	4x rat adrenal glands (0.7 g FW)	II.9.2 (HClO$_4$)
(average: 38%)	(average: 27%)		

Also, the peak shape was improved by increasing the total flow from 0.2 ml/min to 0.5 ml/min and injecting only 10 µl of the sample instead of 60 µl. The gradient was changed to retain morphine longer on the Phenomenex HPLC column (Gemini C18 octadecylsilane, 5 µm, 150 mm × 2 mm combined with a C18 guard column, Phenomenex, 4x2 mm) and to also remove possible interfering molecules causing ion suppression. The final gradient started with 0% B for 4 min, was increased to 60% B over 11 min and finally to 100% B over 1 min. Elution was continued for 2 min at 100% B followed by a 3-min equilibration with starting conditions. Under the optimized conditions the recovery of [N-methyl-CD$_3$]-morphine was improved to 38% for SPE-treated samples without biological background and 27% for [N-methyl-CD$_3$]-morphine in the presence of mouse and rat tissue such as brain and adrenal glands (Tab. 4).

With these optimized LC-conditions it was further investigated if the solvent causes ion suppression and if the MS-signal of SPE eluted morphine could be increased by shifting the retention time of the analyte. For that, a standard solution of 1.74 nmol [N-methyl- CD$_3$]-morphine (500 ng) in 1 ml methanol was prepared and directly infused into the MS with a flow rate of 10 µl/min. The LC-binary gradient was started without injecting a sample into the HPLC-column to observe the ion suppression that is caused by the solvent. A constant signal with an intensity of 20000 cps was observed for [N-methyl-CD$_3$]-morphine (Fig. 7a).

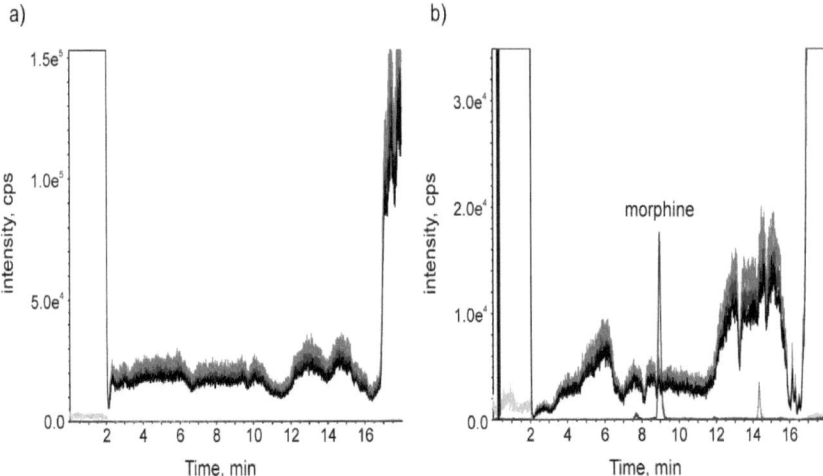

Fig. 7: Ion suppression of morphine as detected by LC-MS.

a) Response of 1.74 nmol [N-methyl- CD_3]-morphine in 1 ml methanol (directly infused into the MS at 10 µl/min) after starting the LC-binary gradient without injecting a sample into the Phenomenex HPLC-column. A constant signal with an intensity of 20000 cps can be observed for [N-methyl- CD_3]-morphine as shown with the black and grey curve.

b) Response of 1.74 nmol [N-methyl- CD_3]-morphine in 1 ml methanol (directly infused into the MS at 10 µl/min) after starting the LC-binary gradient and injection of a mouse brain sample into the HPLC-column that contained unlabeled morphine. A signal for [N-methyl- CD_3]-morphine with an intensity of 4000 cps can be observed when unlabeled morphine elutes in a peak around 9 min.

Interestingly, after 17 min of gradient, when the column is eluted with high organic solvent B (acetonitrile containing 0.1% acetic acid), the signal increased about 5 times. The same experimental set up was conducted only with the difference that a biological sample was injected into the HPLC-column. This sample was obtained after SPE of mouse brain tissue (II.9.2) and contained unlabeled morphine to follow the retention of the analyte. Initially, the signal drastically dropped to an intensity of 1000 cps and stabilized during 8 min of gradient to an intensity of 4000 cps (5 times less than in Fig. 7a) that was constant around 9 min, the time when morphine eluted from the HPLC-column, and increased later on in the gradient to an intensity of 13000 cps (Fig. 7b). The herein observed signal drop of 5 times around 9 min (elution of morphine) explained the low recoveries of 27% after optimization of the LC- method that was solely caused by ion suppression. Further method development on the LC- gradient was declined since a shift of the retention time of morphine would not have led to a decrease of ion suppression.

The influence of the solvent on the intensity of the MS-signal was analyzed to explain whether the increase of MS-signal for [N-methyl-CD$_3$]-morphine 17 min after the gradient was started (Fig. 7a) was caused by the change of ion suppressing solvent A (10% methanol containing 10 mM ammonium acetate and 10 mM ammonium hydroxide) to solvent B (0.1% acetic acid in acetonitrile) during the LC-gradient. Standard solutions of 0.35 nmol [N-methyl-CD$_3$]-morphine (100 ng) in 10% acetonitrile containing different additives were prepared and directly injected into the MS. The highest signal for the product ion m/z 289 for [N-methyl-CD$_3$]-morphine was obtained with 0.1% acetic acid and 0.05% formic acid in 10% acetonitrile (Fig. 8). A drop of MS-signal was observed especially under basic conditions or when ammonium acetate or ammonium formate were added to the solvent.

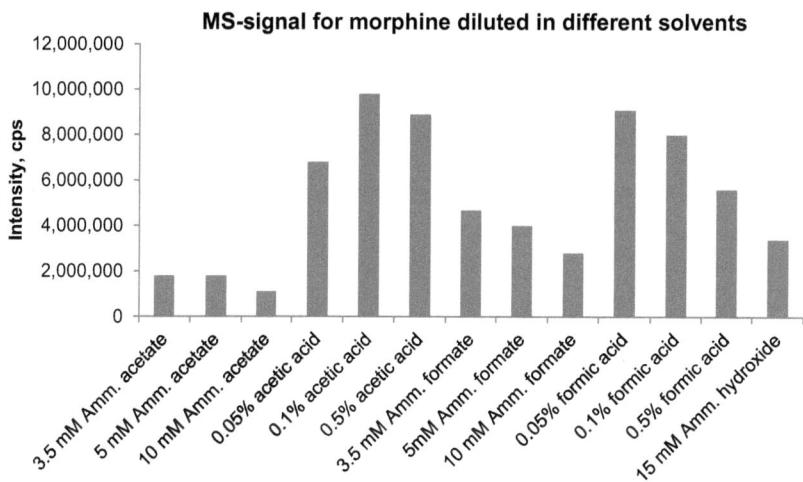

Fig. 8: Intensity of the MS-signal for the product ion m/z 289 of [N-methyl-CD$_3$]-morphine is dependent on the solvent and its additives. Amm. = Ammonium.

Since basic conditions led to a decrease of MS-signal a change to acidic LC-conditions was anticipated. That change led to the formation of two peaks for morphine due to possible ionic interactions of the C18 matrix of the Gemini HPLC column with two differently charged ions of morphine. The Eclipse XDB-C18 HPLC column (Agilent) instead eluted under acidic conditions (II.10.4.A) morphine in only one peak due to exclusive hydrophobic interactions of the C18 matrix with the analyte. However, the use of the Agilent HPLC-column combined to different SPE methods could not further increase the recovery of [N-methyl-CD$_3$]-morphine as shown in Tab. 5.

Tab. 5: Recovery of [N-methyl-CD$_3$]-morphine obtained for differently SPE-treated samples by LC-MS with an Eclipse XDB-C18 HPLC column (II.10.4.A).

cartridge for SPE	SPE-method	recovery	biological background
Strata X-C (Phenomenex)	II.9.1 (TCA)	27% (29%)*	1 rat brain (1.4 g FW)
Strata X-C (Phenomenex)	II.9.2 (HClO$_4$)	18% (39%)*	1 rat brain (1.8 g FW)
Bond Elut Certify (Varian)	II.9.3 (HOFMANN et al. 1999)	21% (50%)* 25% (22%)*	1 rat brain (1.8 g FW) 5 mouse brains (2.1 g FW)
Bond Elut Certify (Varian)	GUARNA et al. 1998, NERI et al. 2008	2% (7%)*	5 mouse brains (2.1 g FW)
Sep Pak C18 (Waters)	II.9.4	3% (13%)*	1 mouse brain (0.4 g FW)
Sep Pak C18 (Waters)	GOUMON et al. 2001	8% (5%)*	6 mouse brains (2.5 g FW)

(%)* Recovery for [N-methyl-CD$_3$]-morphine without biological background

4 MS-Instruments and Sensitivity

Because of the low recovery of morphine with the 4000 QTRAP system solely caused by ion suppression (III.A.3) it was not possible to detect naturally occurring alkaloids in animal tissue or body fluids. Another mass spectrometer available for our analysis was an LTQ-Orbitrap (Thermo Scientific) at the Institute of Environmental Research in the group of Prof. M. Spiteller, Dortmund, Germany. With the LTQ-Orbitrap the SPE-method with Bond Elut Certify cartridges (Varian) as described in II.9.3 was 100% validated and gave a 80-95% recovery of morphine (unpublished, personal communication M. LAMSHÖFT, Dortmund, Germany) whereas with the 4000 QTRAP system the same SPE method gave a recovery of only 21-50% (Tab. 5). In the following the fundamental difference between both MS instruments is explained in detail.

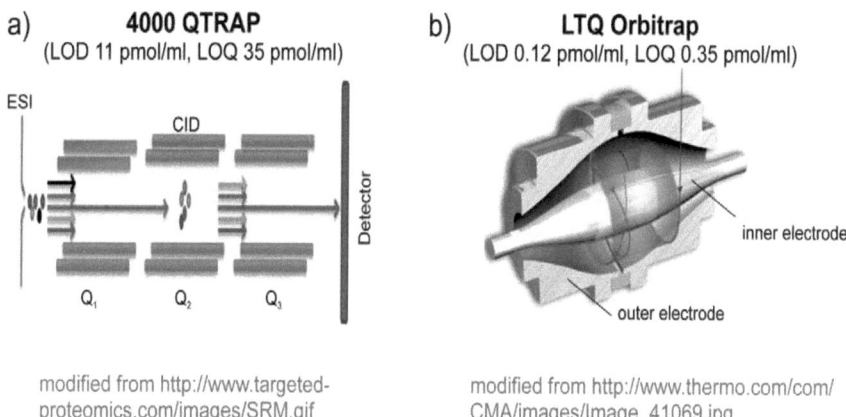

Fig. 9: The principle of ion trapping with the 4000 QTRAP and LTQ-Orbitrap.
a) Ion trapping of the 4000 QTRAP MS instrument.
b) Ion trapping of the LTQ-Orbitrap MS instrument.

The 4000 QTRAP system consists of three quadrupoles (Q1, Q2 and Q3) that are aligned in linear series. Ions travel down the ion path through the electric field generated by the four parallel metal rods of each quadrupole to the detector (electron multiplier) (Fig. 9a). Two different types of scan modes were used to obtain MS data: product ion scan (EPI) and multiple reaction monitoring (MRM). For EPI the analyte is selected by Q1, fragmented in Q2 and fragment ions from the parent ion of the selected analyte are trapped and scanned in Q3 that functions as a linear ion trap. For the more sensitive and specific MRM mode, selected ions for one or more analytes are scanned in Q1, fragmented in Q2 and scanned in Q3 for fragment ions originating from the selected ions in Q1 (so called MRM transitions). In MRM mode both, Q1 and Q3, function as a mass filter in narrow band (unit mass), that results in a reduction of interferences with isobaric masses and increases the signal-to-noise ratio. The higher duty cycle in MRM mode improves the sensitivity in complex matrices.

Ion trapping with the LTQ-Orbitrap is principally different from the 4000 QTRAP. The LTQ-Orbitrap instrument consists of a linear trap quadrupole (LTQ) for the isolation and fragmentation of ions and a C-trap that facilitates ejection of ions into the Orbitrap, which is the actual ion trap, consisting of an inner and outer electrode that generate an electric field. Electrostatic attracted ions enter tangentially to the central electrode, cycle around and along the inner electrode (axial oscillation) and form a rotating ring with a specific frequency that is dependent on the mass of the ion (ion trapping) (Fig. 9b). The frequency of each ring is inferred by measuring the current induced on the outer electrode, and converting the data by Fourier-Transformation to an evaluable MS-signal. The LTQ-Orbitrap is capable of obtaining accurate mass measurements of unknown compounds with high resolution (60000), high mass accuracy (≤ 2 ppm deviation), and high sensitivity (sub femtomolar range). While good quality full scan spectra can be obtained, MS/MS measurement lose sensitivity of up to 100 times. The MS full scan should therefore be used for quantitation in High Resolution mode.

Part B: I.p. Injections into Mice and the Morphine Pathway in Mammals

1 The Biogenesis of Mammalian Morphinan Alkaloids

In 1910, WINTERSTEIN & TRIER predicted that in plants morphinan alkaloids could be biogenetically derived from simple benzylisoquinoline alkaloids (Fig. 10). Almost 50 years later the biogenetic hypothesis was picked up by BARTON & COHEN (1957) postulating that norlaudanosoline is the biogenetic precursor for morphinan alkaloids. This tetrahydroxylated simple benzylisoquinoline also known more commonly in medicine by its other name tetrahydropapaveroline was first isolated in 1909 by PYMAN as a sparingly water-soluble, readily oxidizable reduction product from papaverine that blackens in alkaline solution when left in contact with air. The newfound alkaloid was subsequently tested for its physiological effects in animals (cats, rabbits, dogs) and was reported to relax parts of the smooth musculature and to increase the heart rate caused by a fall of blood pressure when 1-20 mg norlaudanosoline were administered to the animals or animal organs (LAIDLAW 1910). The hypothesis that norlaudanosoline is the ultimate precursor for morphinan alkaloids in plants was seemingly verified by BATTERSBY and his co-workers when they showed that synthesized, radio-labeled norlaudanosoline was successfully incorporated into morphine precursors and morphine in opium poppy and that degradation of the so derived radio-labeled morphinan alkaloids proved position specific incorporation (BATTERSBY & BINKS 1960, BATTERSBY et al. 1963a, BATTERSBY et al. 1964). From then on it was generally accepted that norlaudanosoline is the first intermediate in morphine biosynthesis in plants (CORDELL 1981)

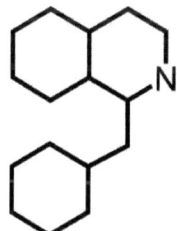
Benzylisoquinoline Alkaloid

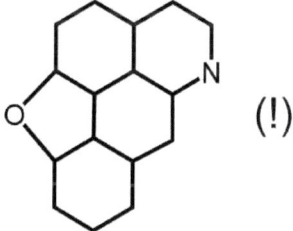
"Suggested Morphine Structure" (1910)

Fig. 10: Biogenetic relationship between benzylisoquinoline and morphinan alkaloids as predicted in WINTERSTEIN & TRIER (1910). Their biogenetic hypothesis is underlain by a wrongly assigned structure of morphinan alkaloids that was later corrected by GULLAND & ROBINSON (1925).

even though tracer experiments on *Papaver somniferum* indicated that norlaudanosoline is less efficiently incorporated into morphine than norcoclaurine, another possible direct precursor (BATTERSBY et al. 1975).The pathway of morphine in opium poppy was eventually revised by ZENK and his co-workers with the discovery that the central intermediate and natural precursor is indeed the trihydroxylated benzylisoquinoline alkaloid norcoclaurine and not the tetraoxygenated norlaudanosoline (STADLER et al. 1989, STADLER & ZENK 1990). Relaxed substrate specificity was found to be the reason for enzymes involved in the early steps of plant morphine biosynthesis to also accept non-natural substrates such as norlaudanosoline. Simultaneous to this discovery, norlaudanosoline was ironically found to occur naturally and after ethanol or L-DOPA administration in animal organs and body fluids (Tab. 6). Studies, focused on the presence of norlaudanosoline after ethanol administration, were based on the idea that an endogenous formation of the alkaloid could be stimulated by ethanol and that the conversion of norlaudanosoline into more complex alkaloids including

Tab. 6: Detection of norlaudanosoline in animals.

Occurrence	Amount Norlaudanosoline	Administered Compound (Dose)	Reference
Human urine	93-930 pmol/ml*	L-DOPA (3-4 g/day)	SANDLER et al. (1973)
Rat brain	4-31 pmol/g FW*	L-DOPA (4 mg/ml drinking water)	TURNER et al. (1974)
Rat brain	0.4 pmol/g FW	L-DOPA (200 mg/kg)	CASHAW et al. (1987)
Rat brain	not quantitated	L-DOPA (200 mg/kg) and Ethanol (3 g/kg)	GERAGHTY & CASHAW (1989)
Human urine	7 pmol/ml	n.a.	MATSUBARA et al. (1992)
	31 pmol/ml	L-DOPA (0.2-0.5 g/day)	
Human urine	0.6-1.1 pmol/ml*	L-DOPA (0.25-1 g/day)	CASHAW (1993b)
Rat brain	0.3 pmol/g FW	Ethanol (3 g/kg)	CASHAW (1993a)
Rat brain	84 pmol/g protein* ((S)-enantiomer)	Ethanol (2 g/kg)	HABER et al. (1997)
Rat brain	7.4 pmol/ml*	n.a.	SÄLLSTRÖM BAUM et al. (1999)
Human brain	0.2 pmol/g FW ((S)-enantiomer)	n.a.	SANGO et al. (2000)

* values found in literature, converted into pmol/ml or pmol/g; n.a., nothing administered.

morphine could be the biochemical basis for alcohol addiction (DAVIS & WALSH 1970, SEEVERS et al. 1970). It was even attempted to show that injections of norlaudanosoline into brain regions of rats could promote alcohol consumption (MYERS & MELCHIOR 1977) leading to various speculations about the alkaloid's involvement in alcoholism (reviewed in COLLINS 2004), an appealing but controversial hypothesis that could not be substantiated.

Increased levels of norlaudanosoline found in organs and body fluids of animals after L-DOPA administration seemed to be plausible since L-DOPA can be converted in humans into dopamine by the action of aromatic L-amino acid decarboxylase (EC 4.1.1.28, ICHINOSE et al. 1989). In a Pictet-Spengler type mechanism, a reaction that had already been predicted in the beginning of the 20th century by WINTERSTEIN & TRIER (1910) and ROBINSON (1917), dopamine then condensates with DOPAL, the aldehyde produced by the action of monoamine oxidase on dopamine which was experimentally proven *in vitro* by HOLTZ et al. (1964) and DAVIS & WALSH (1970). Interestingly, not only increased levels of norlaudanosoline but also codeine and morphine were detected in urine of Parkinson patients undergoing an L-DOPA therapy (MATSUBARA et al. 1992). The researchers suggested that morphine biosynthesis could occur in animals from the distant precursor L-DOPA. A proof for this hypothesis can only come by administration of heavy-isotope labeled L-DOPA which was missed during these studies so that exogenous (dietary) introduction as source for the detected alkaloids could not be excluded. In an attempt to prove the hypothesis that THBIQ and morphine alkaloids found in animals are build *in vivo* from L-DOPA, application of [ring-$^{13}C_6$]-L-DOPA into mice were conducted in this thesis but resulted in total scrambling of the [ring-$^{13}C_6$]-L-DOPA molecule in the animal following i.p. injection.

Norlaudanosoline as possible precursor for mammalian morphine was discussed by several researchers including DAVIS & WALSH (1970), TURNER et al. (1974), CASHAW (1993a) and SANGO et al. (2000) suggesting that a sequence of methylation steps on norlaudanosoline could be followed by an oxidative phenol-coupling step to form the first morphinan precursor salutaridine, the key reaction in morphine biosynthesis. With the goal to verify for the first time the biogenetic hypothesis in animals ZENK and co-workers studied human SH-SY5Y and DAN-G cells that were shown to produce morphine and morphine precursors. After incubation in $^{18}O_2$ atmosphere position-specific $^{18}O_2$-labeled norlaudanosoline, reticuline and morphine were isolated from these cells (POEAKNAPO et al. 2004). A classical precursor feeding experiment with SH-SY5Y cells allowed BOETTCHER et al. (2005) to propose a pathway of morphine in humans starting from two molecules of L-tyrosine *via* the formation of (*S*)-norlaudanosoline with the terminal steps being similar to the plant biosynthesis (Fig. 2). These experiments demonstrated that norlaudanosoline which was previously shown to be built and present in animal tissue (Tab. 6), is the central intermediate in morphine biosynthesis. Based on these facts, i.p. injections into mice described in this thesis were started with the application of norlaudanosoline. The urine of i.p. injected mice was collected *via* a metabolic cage and was scanned for excreted metabolites. Because of our experience with low recoveries and sensitivities with the 4000 QTRAP MS instrument which was caused by ion suppression (III.A.3) and the opportunity to exploit the selectivity, sensitivity and mass accuracy of an LTQ-Orbitrap HR-MS instrument we sent urine samples of injected mice to our collaborating institute. Work-up of the samples was either conducted by our group in St. Louis, USA, or Professor Spiteller's group in Dortmund, Germany.

1.1 Intraperitoneal Injection of (*R*,*S*)-Norlaudanosoline into Mice

Previous experiments with rats that were i.p. injected with norlaudanosoline reported the urinary excretion of the tetrahydroprotoberberine alkaloids coreximine, tetrahydroxyberbine and monomethylated tetrahydroxyberbine but formation of the morphinan alkaloid salutaridine, the key reaction in morphine biosynthesis, was not described (CASHAW *et al.* 1974). Differential enzymatic O-methylated patterns of norlaudanosoline were observed in the brain after intracerebroventricular administration of the alkaloid or after incubation of the alkaloid with soluble rat liver protein but, again, the presence of salutaridine was not reported (MEYERSON *et al.* 1979, CASHAW *et al.* 1983). Experiments with rat liver microsomes and the ultimate precursor of the phenol-coupling reaction reticuline showed that next to the tetrahydroprotoberberine alkaloids coreximine and scoulerine the aporphine isoboldine and the morphinan pallidine were also formed with the latter two suggested to be products of a phenol-coupling reaction in animals, however, salutaridine was not found (KAMETANI *et al.* 1977, 1980). Because a direct morphine precursor has never been found in animals after administration of THBIQ alkaloids the question remained if the living animal is capable of producing morphine precursors or morphine from norlaudanosoline *via* the phenol-coupling reaction. This question was addressed within this thesis by i.p. injection of mice with the early predicted central intermediate followed by urine collection and subsequent isolation and analysis of the metabolites by HR-MS.

Precursor application experiments were started with an i.p. injection of 100 nmol (28.7 µg) (*R*,*S*)-norlaudanosoline*HBr (Acros cat. 221625000) in 200 µl water that was followed by

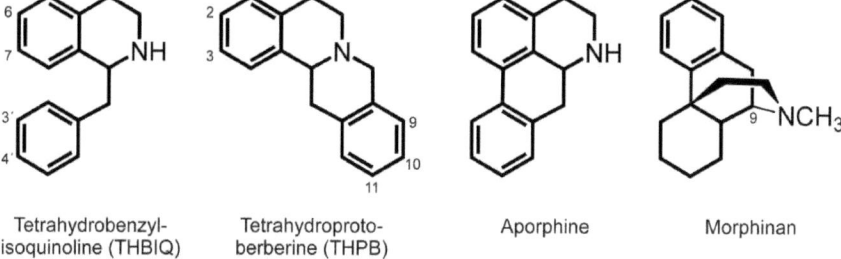

| Tetrahydrobenzyl-isoquinoline (THBIQ) | Tetrahydroproto-berberine (THPB) | Aporphine | Morphinan |

Fig. 11: Types of alkaloids that were found to be excreted after application of the THBIQ alkaloid (*R,S*)-norlaudanosoline to C57 mice.

three additional i.p. injections of 2000 nmol (574 µg) (*R,S*)-norlaudanosoline*HBr in 200 µl water each into two mice every 24 h over 4 days resulting in a total amount of 12.2 µmol (3.5 mg) (*R,S*)-norlaudanosoline*HBr that were administered to two mice (each mouse received in average 61 µmol/kg or 17.5 mg/kg each day). Injected mice were put into a metabolic cage and urine was collected every 24 h and stored at -20°C. A total amount of 6.5 ml urine was collected over 4 days from the two injected mice. Work-up of urine samples of control and injected mice was started by hydrolysis in 2 N HCl for 40 min at 110°C. Additionally, aliquots of urine from control and injected mice were not subjected to hydrolysis to analyze if metabolite profiles are different between both procedures. Subsequently, after adjusting the pH to 7-8, a SPE with a Bond Elut Certify cartridge (Varian) followed and alkaloids that bound to the preconditioned cartridge were first rinsed and finally eluted under basic conditions (II.9.3). HR-LC-MS analysis was done as described in II.10.5. The multitude of metabolites that were found to be excreted was identified by comparison with retention time and fragmentation pattern of reference substances. If standards were unavailable assignment of structures was predicted according to MS/MS characteristics. Postulated metabolites were quantitated with standards exhibiting a similar fragmentation pattern and thus were concluded to be of the same type of alkaloid. Structural assignments for identified metabolites revealed four different types of alkaloids that were detected in the urine of mice i.p. injected with (*R,S*)-norlaudanosoline: tetrahydrobenzylisoquinoline, tetrahydroprotoberberine, aporphine and morphinan alkaloids (Fig. 11). A complete metabolic profile for the identified urinary products carrying the masses *m/z* 288 (4%), 300 (3%), 302 (49%), 314 (10%), 316 (32%), 328 (0.5%) and 330 (1.5%) that were excreted by mice injected with (*R,S*)-norlaudanosoline but not by control mice is shown in Fig. 12. The average recovery of excreted metabolites in urine of mice after (*R,S*)-norlaudanosoline administration that was treated without hydrolysis prior

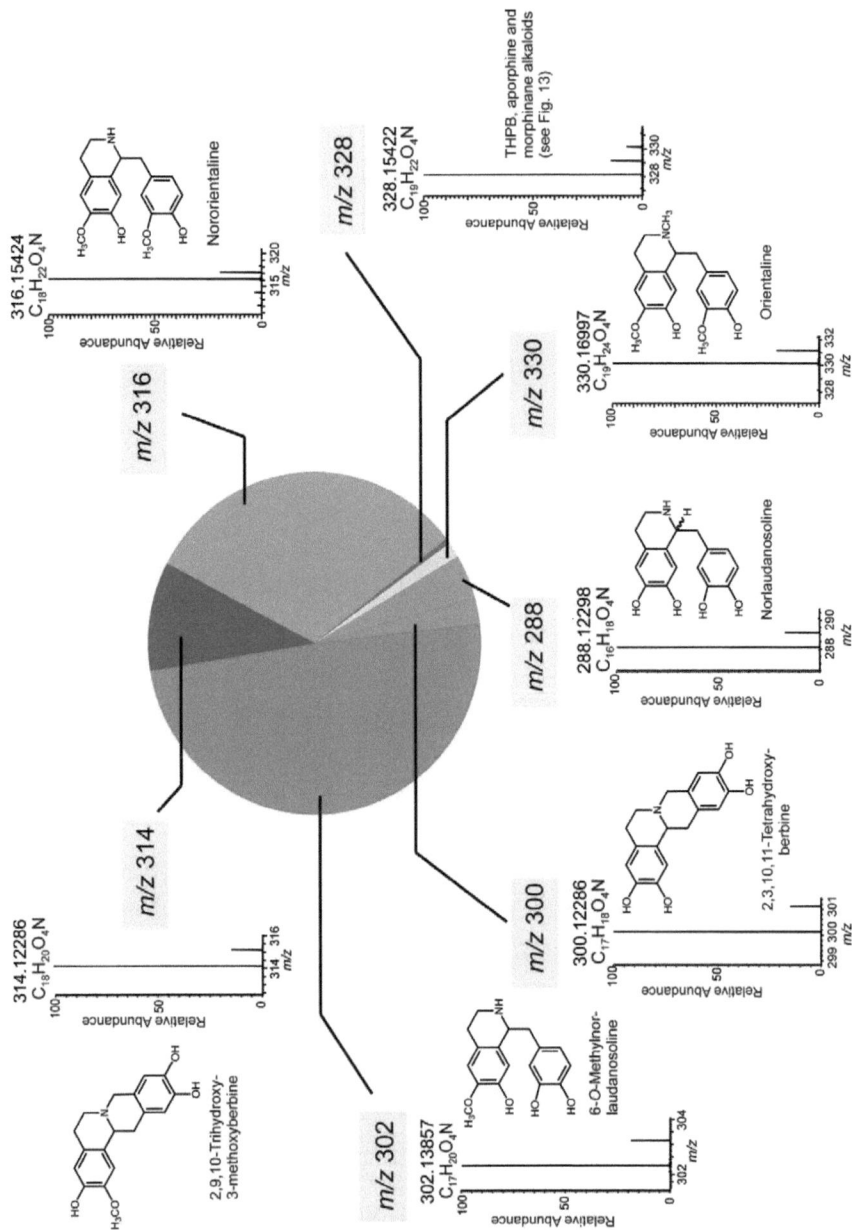

Fig. 12: Metabolites found in the urine of mice i.p. injected with (R,S)-norlaudanosoline (0.1 μmol/24h for 1 day followed by 2 μmol/24h for 3 days). The HR-LC-MS analysis revealed the urinary excretion of alkaloids from the THBIQ, tetrahydroprotoberberine, aporphine and morphinan type.

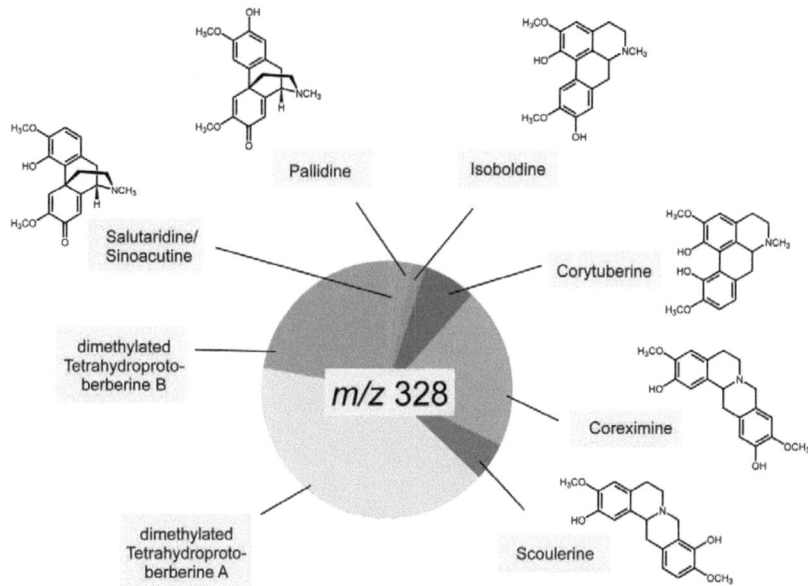

Fig. 13: Metabolites with a mass of m/z 328 found in the urine of mice i.p. injected with (R,S)-norlaudanosoline (0.1 µmol/24 h for 1 day followed by 2 µmol/24 h for 4 days). The HR-LC-MS analysis revealed the urinary excretion of four tetrahydroprotoberberines including coreximine and scoulerine, two morphinan alkaloids salutaridine and pallidine as well as two aporphine alkaloids isoboldine and corytuberine.

to SPE was only about 2%. Generally, hydrolysis was found to increase the recovery of excreted metabolites, particularly m/z 288 and m/z 302, to 20-40%. Because of its complexity and importance a detailed metabolic profile for the ions with m/z 328 excreted by mice i.p. injected with (R,S)-norlaudanosoline is shown in Fig. 13.

1.1.1 Urinary Excretion of Tetrahydrobenzylisoquinoline Alkaloids after Application of (R,S)-Norlaudanosoline

Metabolites belonging to the THBIQ type of alkaloids were represented by the masses m/z 288, m/z 302, m/z 316 and m/z 330. The ion m/z 288 was identified as norlaudanosoline and 4% of the total amount of i.p injected (R,S)-norlaudanosoline was recovered unmetabolized from urine of i.p. injected mice. The other main metabolites with m/z 302 and m/z 316 as well as the minor metabolites with m/z 330 were identified as mono-, di- and trimethylated norlaudanosoline derivatives, respectively. The *in vitro* conversion of norlaudanosoline to monomethylated norlaudanosoline derivatives had already been described in studies with rat liver by MEYERSON et al (1979). Readily detectable amounts of 6-,7- and 3'-O-methylated norlaudanosoline were isolated from rat brain after intracerebroventricular injection of (R)- or (S)-norlaudanosoline (CASHAW et al. 1983). Because of unstable retention times during HR-LC-MS analysis, non-optimal separation conditions for these hydrophilic alkaloids and limited availability to reference substances a clear assignment to either one of the structural isomers of monomethylated THBIQ alkaloids could not be achieved in this thesis. But based on MS/MS data it can be postulated that most likely 6- and 7-O-methylated norlaudanosoline were excreted as well as one norlaudanosoline carrying a methoxy group on either C-3' or C-4' position (for localizing positions 3´, 4´, 6 and 7 in the THBIQ molecule see Fig. 11). A comparison with standards revealed that the ion m/z 316 represented nororientaline, a 6,3'-O-dimethylated norlaudanosoline derivative, rather than norisoorientaline, the 7,3'-O-dimethylated norlaudanosoline derivative. A clear statement is difficult since both structural isomers could not be base-line separated under the used LC-separation conditions. A mixture of four metabolites carrying the mass m/z 330 was identified and assigned to THBIQ type of alkaloids carrying three methyl groups. However, comparison with the only available reference, reticuline (m/z 330) did not superimpose with any of these four detected metabolites leaving a total of five other possible THBIQ structures with m/z 330. Orientaline (Fig. 12) is an example of one of the possible predictions for m/z 330.

1.1.2 Urinary Excretion of Tetrahydroprotoberberine Alkaloids after Application of (R,S)-Norlaudanosoline

The two metabolites with m/z 300 and three metabolites with m/z 314 that were found in mice after i.p. injection of (R,S)-norlaudanosoline were predicted to be alkaloids of the tetrahydroprotoberberine type. Based on MS/MS data and previous findings by CASHAW et al. (1974) the two ions with m/z 300 were predicted as 2,3,9,10-tetrahydroxyberbine and 2,3,10,11-tetrahydroxyberbine (for localizing positions 2, 3, 9, 10 and 11 in the molecule see Fig. 11). Their monomethylated derivatives (plus one methyl group = 14 mass units), e.g. 2,10,11-trihydroxy-3-methoxyberbine, were represented by the three ions with m/z 314.

A second addition of a methyl group (+14) leads to mass m/z 328 and, indeed, at least four out of eight metabolites with a mass of m/z 328 represented alkaloids from the tetrahydroprotoberberine type (Fig. 13). Two metabolites (tetrahydroprotoberberine A and tetrahydroprotoberberine B) could not be further identified, but a comparison of the remaining two ions with standards revealed scoulerine and coreximine as metabolic products of i.p. administered (R,S)-norlaudanosoline*HBr to mice. The presence of coreximine in urine of rats after i.p. injection of 189 mg norlaudanosoline*HBr/kg had also been described by CASHAW et al. (1974). Both tetrahydroprotoberberine alkaloids, scoulerine and coreximine, were found by KAMETANI et al. (1977, 1980) as products from the incubation of reticuline with rat liver microsomes.

1.1.3 Urinary Excretion of Aporphine and Morphinan Alkaloids after Application of (R,S)-Norlaudanosoline

A most interesting group of metabolites excreted by mice after (R,S)-norlaudanosoline application contained four ions exhibiting all a mass of m/z 328 that were found to belong to the two types of aporphine and morphinan alkaloids. Comparison with standards revealed the urinary excretion of corytuberine (aporphine), isoboldine (aporphine), pallidine (morphinan) and, for the first time, salutaridine/sinoacutine (morphinan). Enantiomers of corytuberine, isoboldine and pallidine are distinguished by their optical rotations and therefore labeled as (+) or (-). The optical antipode of salutaridine is called sinoacutine that does not serve as a morphine precursor. The urinary excretion of the morphinan alkaloid salutaridine/sinoacutine after administration of the distant precursor (R,S)-norlaudanosoline has never been reported before and seems to verify the biogenetic hypothesis which was predicted earlier in 1957 by BARTON & COHEN and that suggested norlaudanosoline as morphine precursor. The presence of all four metabolites corytuberine, isoboldine, pallidine and salutaridine/sinoacutine is proof that the phenol-coupling, the key reaction of morphine biosynthesis, takes place in animals. The oxidative C-C-coupling of reticuline in animals leading to this set of four products is explained in detail later in this thesis when various mammalian enzymes catalyzing the phenol-coupling reaction are introduced (see III.C.4). Retention time, HR-MS and MS/MS data of a salutaridine standard confirmed unequivocally the presence of salutaridine/sinoacutine in the urine of mice i.p. injected with (R,S)-norlaudanosoline (Fig. 14). The morpinandienone alkaloid was not found to be present in control urine or in the application solution, but was detected in a concentration of 12-31 pmol/ml urine (4-10 ng/ml urine) of mice i.p. injected with (R,S)-norlaudanosoline. Because of these low amounts of alkaloid up to three SPE-treated urine samples had to be combined and concentrated to obtain good quality MS/MS signals (Fig. 14b). As an alternative a standard addition experiment was conducted to unequivocally confirm the presence of the morphinan precursor in urine of i.p. injected mice. For this experiment, SPE-treated urine samples of i.p. injected mice found to contain 1-3 pmol (0.3-1 ng) salutaridine/sinoacutine were mixed with ca. 1.5 pmol (0.5 ng) salutaridine standard and analyzed by HR-LC-MS. The analysis revealed a proportional increase of the peak area corresponding to salutaridine after addition of the standard.

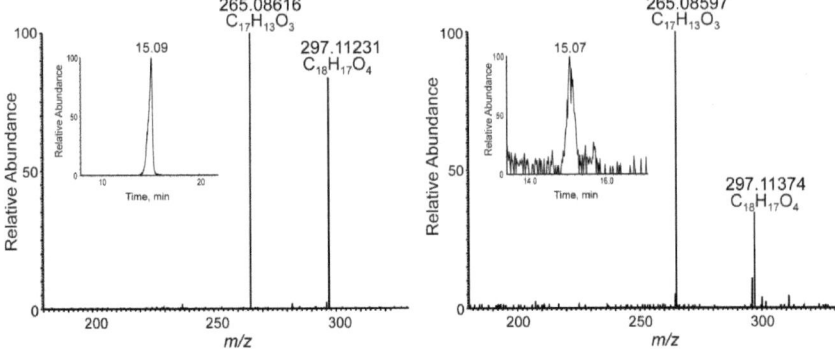

Fig. 14: Comparison of salutaridine standard and salutaridine/sinoacutine excreted by mice i.p. injected with (R,S)-norlaudanosoline (0.1 µmol/24 h for 1 day followed by 2 µmol/24 h for 4 days).
a) Full scan and HR-MS spectra of salutaridine standard.
b) Full scan and HR-MS spectra of salutaridine/sinoacutine detected in urine of mice after administration of (R,S)-norlaudanosoline.

1.2 Intraperitoneal Injection of (R)-Norlaudanosoline into Mice

The morphinan alkaloid salutaridine/sinoacutine that was excreted after administration of (R,S)-norlaudanosoline was further analyzed for the right stereochemistry to serve as a precursors for morphine. Since a chiral separation of salutaridine and its enantiomer sinoacutine was not accomplished a different approach was chosen based on the discovery of a human (R)-specific enzyme that was found to N-methylate small molecules like THBIQ alkaloids and is described in detail later in this thesis (III.C.3). The hypothesis that the early steps in mammalian morphine biosynthesis could all be (R)-configurated was attempted to be proven by i.p. injection of chemically synthesized (R)-norlaudanosoline into mice.

A chemical synthesis was aimed to the production of (R)-norlaudanosoline*HBr. Treatment of (R)-norreticuline that was obtained from the departmental collection with*HBr under heating for 2 hrs following further purification by preparative HPLC under the conditions described in II.10.3 gave the trifluoroacetate of (R)-norlaudanosoline (RICE & BROSSI 1980, see II.5.2). The hydrobromide of (R)-norlaudanosoline was prepared according to II.5.2 by loading an aliquot of the trifluoroacetate onto TLC. The UV-corresponding band was scratched off, eluted with methanol, evaporated, resuspended in HBr and lyophilized. (R)-norlaudanosoline*HBr was verified by LC-MS analysis and subsequently administered to mice by starting with a first i.p. injection of 100 nmol (28.7 µg) (R)-norlaudanosoline*HBr in

200 µl water that was followed by three additional i.p. injections of 2000 nmol (574 µg) (*R*)-norlaudanosoline*HBr in 200 µl water each into two black mice every 24 h over 4 days resulting in a total amount of 12.2 µmol (3.5 mg) (*R*)-norlaudanosoline*HBr that were administered to two mice (each mouse received 61 µmol/kg or 17.5 mg/kg each day). Injected mice were put into a metabolic cage and urine was collected every 24 h and stored at -20°C. A total amount of 6.7 ml urine was collected over 4 days from the two injected mice. Urine samples from control and injected mice were subjected to SPE with Bond Elut Certify either with or without previous hydrolysis in 2 N HCL for 40 min at 110°C according to II.9.3. An adjustment of the pH to 7-8 was followed by a SPE with Bond Elut Certify and alkaloids that bound to the preconditioned cartridge were first rinsed and finally eluted under basic conditions. HR-LC-MS analysis was done under the conditions as described in II.10.5. Metabolites were identified by comparison with retention time and fragmentation pattern of standards as described in detail for the application of (*R,S*)-norlaudanosoline.

A similar excretion pattern for the ions *m/z* 288 (3%), 300 (2%), 302 (43%), 314 (2%), 316 (49%), 328 (0.3%) and 330 (0.5%) as for the application of (*R,S*)-norlaudanosoline*HBr was observed when (*R*)-norlaudanosoline*HBr was administered. The metabolites were not detected in the application solutions and control urine. Again, it was shown that on average the recovery of excreted metabolites in urine of mice was lower without hydrolysis (2%) compared to samples that were hydrolyzed (40%). Metabolites that were exclusively found in urine of i.p. injected mice were classified into several types of alkaloids as described earlier. The analysis revealed that the main products carrying the masses *m/z* 288, 300, 302, 314, 316, and 330 belonged to the THBIQ and THPB type of alkaloids and structures could be predicted as already described (III.B.1.1.1 and 1.1.2). Only a minor portion represented six metabolites with a mass of *m/z* 328 of which four of them were identified as alkaloids from the THPB type. Aporphine alkaloids were not found to be present, probably due to concentrations below the limit of detection, but the two morphinan alkaloids pallidine and salutaridine were detected. The occurrence of the morphinan alkaloid salutaridine in a concentration of 3 pmol/ml urine (1 ng/ml urine) was unequivocally confirmed by retention time and HR-MS (Fig. 15) as well as by standard addition experiments.

This finding not only verified again that the living animals is capable of forming morphine precursors from norlaudanosoline but also suggested that the early steps in mammalian morphine biosynthesis occur fundamentally different than the ones in the plant pathway

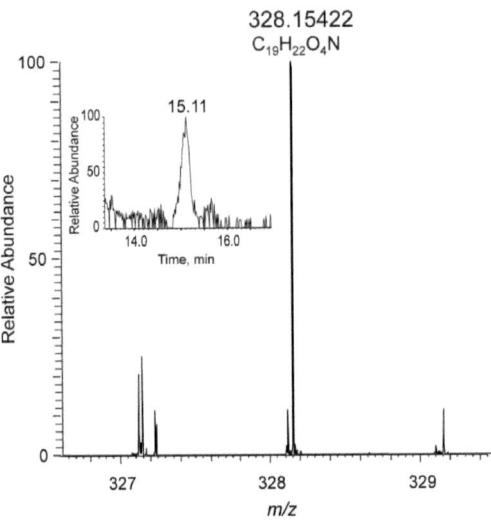

Fig. 15: Full scan and HR-MS spectrum of salutaridine excreted by mice i.p. injected with (R)-norlaudanosoline (0.1 µmol/24 h for 1 day followed by 2 µmol/24 h for 3 days).

provided that the (R)-configuration is not changed *in vivo*. In animals the (R)-enantiomer of the universal mammalian precursor norlaudanosoline was found within this thesis to be converted to salutaridine. In opium poppy, (S)-configurated norcoclaurine is the first central intermediate that is further converted to (S)-reticuline that then undergoes a change of configuration to form (R)-reticuline, the ultimate precursor of the phenol-coupling reaction yielding salutaridine.

1.3 Intraperitoneal Injection of (R)-[N-CD$_3$]-Reticuline into Mice

The key reaction in morphine biosynthesis, in which the benzylisoquinoline precursor reticuline is converted by oxidative C-C-phenol-coupling to the morphinan alkaloid salutaridine, was first proposed in 1957 by Derek H.R. Barton (BARTON & COHEN 1957), who was later awarded the Nobel Prize in 1969 for his "Principles in Conformational Analysis". The two phenolic hydroxy groups of reticuline seemed to be in the right position to build *via* phenol-coupling the morphinan skeleton and indeed, racemic radio-labeled reticuline was fed to opium poppy and incorporated into morphine (BARTON *et al.* 1963a, BATTERSBY *et al.* 1963a, BARTON *et al.* 1965). However, it was demonstrated that only one of the two isomers of reticuline, the (R)-(-)-form, is the direct substrate for morphine in opium poppy suggesting that the enzyme system catalyzing the oxidative phenol-coupling reaction is stereospecific

(BATTERSBY et al. 1965). Recognizing that feeding of the labeled precursor revealed the biosynthesis in opium poppy, WEITZ et al. (1987) proposed that morphine could be possibly synthesized in animals by observing the *in vitro* and *in vivo* conversion of reticuline into salutaridine in rat. Later on, the C-C-phenol-coupling of (*R*)-reticuline to salutaridine was demonstrated with microsomes from pig, beef, sheep and mouse liver. Pig liver showed the highest conversion rate and the organ was chosen for the purification of salutaridine synthase (AMANN et al. 1995).

Since the urinary excretion of salutaridine following i.p. injection of (*R*)-norlaudanosoline*HBr into mice was described in this thesis (III.B.1.2) it was assumed that i.p. injection of (*R*)-reticuline into mice could also result in the urinary excretion of salutaridine and thus may give further evidence for a biosynthesis of morphine in mammals. (*R*)-[*N*-CD$_3$]-reticuline was chosen for the application experiments since it was available from our departmental collection. An endogenous conversion into salutaridine would only be proven if the deuterium label was position-specifically incorporated into salutaridine.

(*R*)-[*N*-CD$_3$]-reticuline was obtained from our departmental collection and *ca.* 50 nmol (15.7 μg) in 200 μl water were i.p. injected each into five mice every 24 h over 4 days resulting in a total amount of 945 nmol (0.314 mg) (*R*)-[*N*-CD$_3$]-reticuline that were administered to five mice (each mouse received *ca.* 2 μmol/kg or 0.628 mg/kg each day). Injected mice were put into a metabolic cage and urine was collected every 24 h and stored at -20°C. A total amount of 12 ml urine was collected over 4 days from the two injected mice. Urine samples from injected mice were subjected to SPE with Bond Elut Certify either with or without previous hydrolysis in 2 N HCL for 40 min at 110°C according to II.9.3. An adjustment of the pH to 7-8 was followed by a SPE with Bond Elut Certify and alkaloids that bound to the preconditioned cartridge were first rinsed and finally eluted under basic conditions. HR-LC-MS analysis was done under the conditions as described in II.10.5. Urinary metabolites were identified by comparison with retention time and HR-MS of standards. Heavy-isotope labeled products that were detected in the urine of injected mice added up to a total of 16% recovered substance of which the main part was identified as unmetabolized (*R*)-[*N*-CD$_3$]-reticuline (12 μmol/ml urine or 4.1 μg/ml urine, about 97.5%). The minor portion contained trace amounts of two heavy-isotope labeled THPB alkaloids coreximine and scoulerine with *m/z* 330.16679 and a deviation from the theoretical value of -0.31960 ppm (Fig. 16a,b).

a) Standards Coreximine and Scoulerine (m/z 328.15426)

b) Metabolites with m/z 330.16679 after i.p.injection of (R)-[N-CD$_3$]-Reticuline

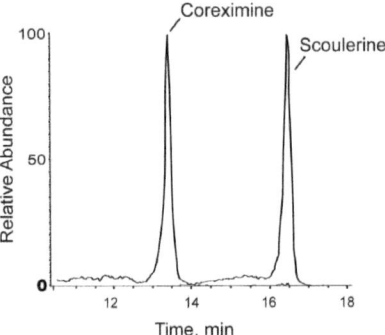
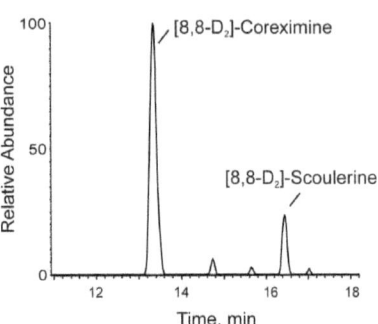

c) Standards Pallidine, Corytuberine, Salutaridine and Isoboldine (m/z 328.15423)

d) Metabolites with m/z 331.17326 after i.p.injection of (R)-[N-CD$_3$]-Reticuline

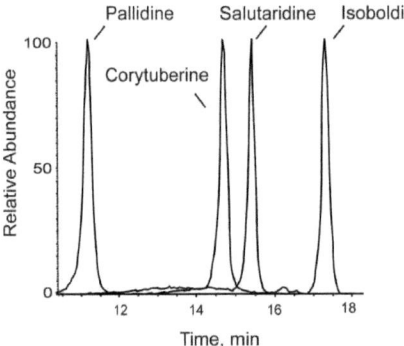
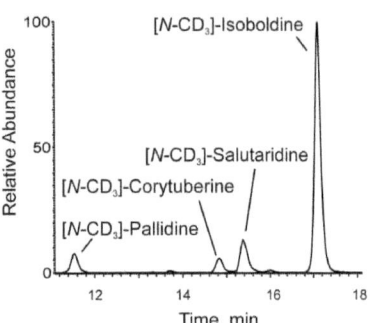

Fig. 16: HR-LC-MS full scan analysis reveals the urinary excretion of heavy-isotope labeled metabolites after i.p. injection of (R)-[N-CD$_3$]-reticuline (50 nmol/24 h for 4 days). The reference substances are set to 100% because they were each separately added to one chromatogram.

a) Full scan extracted chromatogram of the tetrahydroprotoberberine alkaloid standards coreximine and scoulerine with a mass of m/z 328.15426 and chemical formula of C$_{19}$H$_{22}$O$_4$N.
b) Full scan extracted chromatogram of urinary products verified as heavy-isotope labeled coreximine (2.7 pmol/ml urine or 0.9 ng/ml urine) and scoulerine (0.2 pmol/ml urine or 0.07 ng/ml urine) with a mass of m/z 330.16679 and chemical formula of C$_{19}$H$_{20}^{2}$H$_2$O$_4$N.
c) Full scan extracted chromatogram of the aporphine and morphinan alkaloid standards pallidine, corytuberine, salutaridine and isoboldine with a mass of m/z 328.15423 and a chemical formula of C$_{19}$H$_{22}$O$_4$N.
d) Full scan extracted chromatogram of urinary products identified as heavy-isotope labeled pallidine (15 pmol/ml urine or 5 ng/ml urine), corytuberine (9.1 pmol/ml urine or 3 ng/ml urine), salutaridine (52 pmol/ml urine or 17 ng/ml urine) and isoboldine (242 pmol/ml urine or 80 ng/ml urine) with a mass of m/z 331.17326 and chemical formula of C$_{19}$H$_{19}^{2}$H$_3$O$_4$N.

Three smaller peaks with a signal-to-noise ratio less than 10 also occurred in this full scan chromatogram and were not further identified. Four ions with the mass of m/z 331.17326 and a deviation from the theoretical value of -0.31960 ppm were verified as heavy-isotope labeled salutaridine, pallidine, corytuberine and isoboldine (Fig. 16c,d). HR-MS analysis suggested a chemical formula of $C_{19}H_{19}{}^2H_3O_4N$ ($\leq \pm$ 1 ppm) for the excreted aporphine and morphinan alkaloids which are phenol-coupled products as it will be demonstrated in detail later in this thesis when mammalian phenol-coupling enzymes are introduced (III.C.4). The HR-MS analysis of coreximine and scoulerine proposed a chemical formula of $C_{19}H_{20}{}^2H_2O_4N$ ($\leq \pm$ 1 ppm) indicating that only two deuteriums were incorporated into the THPB alkaloids. Optimal LC-conditions enabled the clear separation of this set of products formed from (R)-$[N$-$CD_3]$-reticuline. With the detection of incorporated deuterium-label in salutaridine, that was formed in a concentration of 52 pmol/ml urine (17 ng/ml urine), it was unequivocally proven that the phenol-coupling reaction occurs endogenously in the animal. The results obtained here in this experiment were very similar to *in vitro* studies on the biotransformation of reticuline in animals demonstrating that incubation of racemic reticuline and (S)-reticuline with rat liver in the presence of NADHP gave four products coreximine, scoulerine, pallidine and isoboldine (KAMETANI *et al.* 1977, 1980). However, in these experiments the phenol-coupled product salutaridine could not be identified and corytuberine escaped detection as well.

The previous chapter III.B.1 described the detection of salutaridine in urine of mice i.p. injected with (R,S)-norlaudanosoline, (R)-norlaudanosoline and (R)-$[N$-$CD_3]$-reticuline thus confirming that animals have the capacity to produce morphine alkaloids from distant precursors. In all application experiments the occurrence of the morphine precursor was verified by HR-MS with a high mass accuracy < 1 ppm (Tab. 7). The *in vivo* studies described herein proved unequivocally and for the first time that the morphinan mammalian alkaloid salutaridine is biogenetically derived from the simple benzylisoquinoline precursor (R)-norlaudanosoline suggesting a fundamental difference from the early steps in the opium poppy pathway. A possible (R)-configurated route from norlaudanosoline to salutaridine will be further substantiated in this thesis at a later point (III.C.3). The presence of a set of four heavy-isotope labeled phenol-coupled products after administration of (R)-$[N$-$CD_3]$-reticuline will be further elucidated in III.C.4 when mammalian P450 enzymes involved in the C-C-phenol-coupling reaction of (R)-reticuline to salutaridine are described.

Tab. 7: A high mass accuracy for salutaridine confirms unequivocally the detection of the morphine precursor in urine of mice i.p. injected with (R,S)-norlaudanosoline, (R)-norlaudanosoline and (R)-[N-CD_3]-reticuline.

I.p. injected compound	m/z Salutaridine, measured	m/z Salutaridine, theoretical	Mass accuracy (ppm)	Concentration (per ml urine)
(R,S)-norlaudanosoline	328.15423	328.15433	-0.31082	12-30 pmol (4-10 ng)
(R)-norlaudanosoline	328.15422	328.15433	-0.35327	3 pmol (1 ng)
(R)-[N-CD_3]-reticuline	331.17326	331.17317	0.28610	52 pmol (17 ng)

2 Thebaine formation from Salutaridine in Mammals

Salutaridine was first chemically synthesized from thebaine by BARTON et al. (1963a). The morphinandienone obtained its name after a direct comparison with the natural alkaloid that was isolated from *Croton salutaris* (BARNES 1964). The *para-ortho* phenol-coupled product of the key reaction in morphine biogenesis had been predicted in 1957 by BARTON & COHEN to be the ultimate precursor of the morphine skeleton. Based on the idea of STORK (1960) the proposed reaction mechanism of the conversion of salutaridine to the first pentacyclic alkaloid thebaine was predicted to proceed *via* a reduction of the dienone salutaridine to its dienol salutaridinol which then undergoes allylic elimination to form thebaine (BATTERSBY 1963). Subsequently, after having salutaridine in hand, two groups in England around BARTON and BATTERSBY aimed to find experimental evidence for the biogenetic proposal by conducting independent and parallel sets of studies. Both research teams agreed that chemically synthesized radio-labeled salutaridine was found to be efficiently incorporated into thebaine, codeine and morphine by opium poppy capsules (BARTON et al. 1965).

Since precursor feeding of salutaridine elucidated morphine biosynthesis in opium poppy, DONNERER et al. (1986) injected rats intravenously into the tail vein with a precursor feeding solution of salutaridine (10 mg/kg body weight) to prove the occurrence of morphine biosynthesis in animals that was assumed to be similar to the one in plants. Indeed, increased amounts of codeine and morphine were detected in several organs isolated from the injected rats suggesting that the same biosynthetical pathway existed as in the poppy plant. The question whether a mammalian enzyme system could be found, as it was shown for opium poppy, responsible for the reduction of salutaridine to salutaridinol was addressed by FISINGER (1998) reporting that cytosolic rat liver protein catalyzed stereoselectively the

NADPH-dependent reaction of radio-labeled salutaridine to salutaridinol that, moreover, could further be converted to thebaine with cytosolic and microsomal rat liver protein. Within this thesis, i.p. injections of the morphinan alkaloids salutaridine and salutaridinol into mice should further demonstrate the capability of the living animal to biosynthesize the pentacyclic morphine skeleton from salutaridine and its alcohol *in vivo*.

2.1 Intraperitoneal Injection of Salutaridine into Mice

As described earlier in this thesis, salutaridine was present in urine of mice after i.p. injection of (*R,S*)-norlaudanosoline, (*R*)-norlaudanosoline and (*R*)-[*N*-CD$_3$]-reticuline (III.B.1) serving as evidence that the living animal is capable to biosynthesize the morphinan dienone from distant precursors *via* the phenol-coupling reaction. The following application experiment on mice was conducted to further elucidate the ability of animals to catalyze the *in vivo* conversion of salutaridine to thebaine and thus the closure of the oxide bridge. For that, 500 nmol (163.5 µg) salutaridine obtained from our departmental collection were i.p. injected into each of two mice every 24 h over 4 days resulting in a total amount of 4000 nmol (1.3 mg) salutaridine that were administered (each mouse received 20 µmol/kg or 6.5 mg/kg each day). Injected mice were put into a metabolic cage and urine was collected every 24 h and stored at -20°C. A total amount of 6 ml urine was collected over 4 days from the two i.p. injected mice. Urine samples from injected mice were adjusted to pH 7-8 without any previous HCl-treatment to avoid non-enzymatically acid catalyzed conversion of possibly formed salutaridinol into thebaine as it has been reported by BARTON *et al.* (1965). SPE with Bond Elut Certify cartridges (Varian) was conducted and alkaloids that bound to the preconditioned cartridge were first rinsed and finally eluted under basic conditions (II.9.3). HR-LC-MS analyses of urine samples were done under the conditions as described in II.10.5. Excreted metabolites formed from salutaridine *in vivo* after i.p. injection into mice were identified by comparison with retention time and HR-MS of standards as salutaridinol and thebaine (Fig. 17). These findings verified that in animals salutaridine, that is biogenetically derived from the ultimate THBIQ precursor norlaudanosoline, will further be transformed into the pentacyclic morphine skeleton.

Additionally, unmetabolized salutaridine and demethylated salutaridine were also found to be present. High-resolution data and amounts of the verified compounds that were neither found in control urine nor in the application solution are listed in Tab. 8. A fifth metabolite with the

a) Salutaridinol standard:
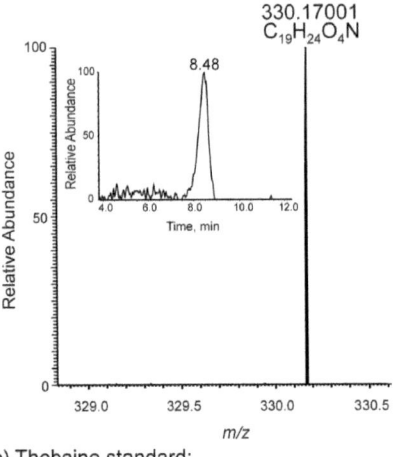

b) Salutaridinol isolated from urine of mice i.p. injected with Salutaridine:
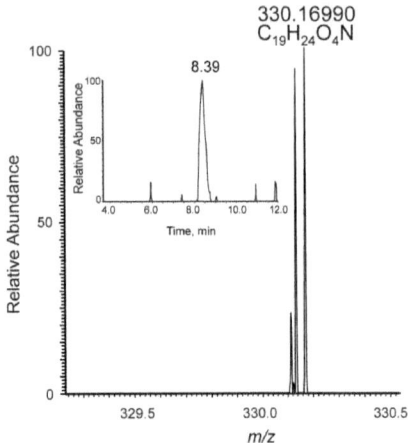

c) Thebaine standard:
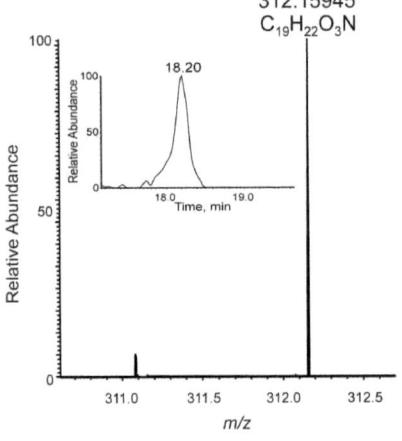

d) Thebaine isolated from urine of mice i.p. injected with Salutaridine:
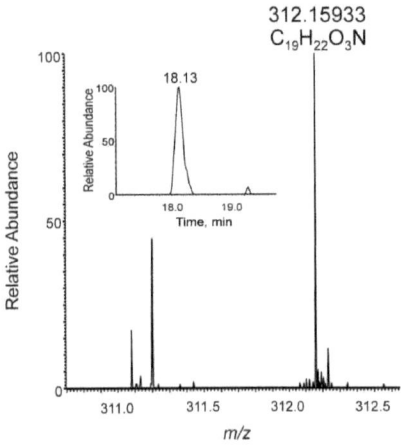

Fig. 17: HR-MS of salutaridinol and thebaine standard and the urinary metabolites salutaridinol and thebaine that were detected in urine of mice i.p. injected with salutaridine (500 nmol/24 h for 4 days).

a) HR-MS product ion and full scan extracted chromatogram of salutaridinol standard.
b) HR-MS product ion and full scan extracted chromatogram of urinary excreted salutaridinol.
c) HR-MS product ion and full scan extracted chromatogram of thebaine standard.
d) HR-MS product ion and full scan extracted chromatogram of urinary excreted thebaine.

Tab. 8: HR-MS data for all metabolites detected in the urine of mice i.p. injected with salutaridine (500 nmol/24 h for 4 days).

Metabolite	Mass measured (m/z)	Mass theoretical (m/z)	Mass accuracy (ppm)	Concentration (per ml urine)
thebaine	312.15933	312.15942	-0.28100	0.6 pmol (0.2 ng)
demethylated salutaridine	314.13861	314.13869	-0.25327	1.1 nmol (330 ng)
salutaridine	328.15427	328.15433	-0.18357	1 nmol (330 ng)
salutaridinol	330.16990	330.16999	-0.13855	3 pmol (1 ng)

same HR-MS as salutaridinol was also found but was excluded to be the epimer *epi*-salutaridinol because it eluted at a different elution time. Because only salutaridinol and not the diastereomer, that is inactive as a precursor for morphine in plants, was detected in the urine of mice i.p. injected with salutaridine, the biotransformation of salutaridine to salutaridinol was therefore concluded to occur stereoselectively confirming the findings of FISINGER (1998). Quantitation of all identified compounds yielded only 0.3% of the total amount of i.p. injected substance that could be recovered in mouse urine.

2.2 Intraperitoneal Injection of [7D]-Salutaridinol into Mice

In vitro studies with Acetyl CoA and rat liver protein showed that radio-labeled thebaine was formed as the only product from radio-labeled salutaridinol at pH 9 (FISINGER 1998). Interestingly, no reaction product was observed with boiled protein or absent cofactor suggesting the presence of an essential enzyme system similar to the one in opium poppy that catalyzes the CoA-dependent formation of the oxide bridge from salutaridinol having been activated before by an ester at the C-7-hydroxy group. It was demonstrated in this thesis that application of salutaridine to mice led to the urinary excretion of salutaridinol and thebaine (III.B.2.1). In the following experiment it was attempted to confirm that salutaridinol is the ultimate precursor for the *in vivo* formation of thebaine. Moreover, i.p. injections with deuterium labeled salutaridinol were expected to lead to the urinary excretion of morphine precursors with position-specific incorporation of deuterium proving an endogenous biosynthesis. For that, [7D]-salutaridinol was synthesized as described in II.5.3 from salutaridine through reduction with sodium borodeuteride. The purity of the chemically synthesized compound was analyzed by HR-MS (II.10.5); the MS is shown in Fig. 3. 50 nmol (16.5 µg) [7D]-salutaridinol were i.p. injected each into two mice every 24 h over 4 days resulting in a total amount of 400 nmol (0.132 mg) [7D]-salutaridinol that were administered (each mouse received 2 µmol/kg or 0.66 mg/kg each day). Injected mice were put into a metabolic cage and urine was collected every 24 h and stored at -20°C. A total amount of 6 ml urine was collected over 4 days from the two i.p. injected mice. As described for the application of salutaridine urine samples from injected mice were adjusted to pH 7-8 without any previous HCl-treatment. That precaution was necessary to avoid a non-enzymatically acid catalyzed transformation into the pentacyclic alkaloid. A SPE with Bond Elut Certify cartridges (Varian) was conducted. Alkaloids that bound to the preconditioned cartridge were first rinsed and finally eluted under basic conditions (II.9.3). HR-LC-MS analyses of urine samples were done under the conditions as described in II.10.5. A comparison of retention time, HR-MS and MS/MS spectra with standards verified that among the metabolites [7D]-thebaine was present and excreted in urine of mice i.p. injected with [7D]-salutaridinol (Fig. 18). The incorporation of the deuterium-label into the pentacyclic alkaloid proved unequivocally and for the first time that the reaction is occurring endogenously in the living animal. In addition to unmetabolized [7D]-salutaridinol, unlabeled salutaridine, salutaridinol and thebaine as well as demethylated salutaridine were also found to be urinary products (Tab. 9) adding up to a total of 1.2% of i.p. injected substance recovered as metabolites in the urine. It was shown for the opium poppy enzyme salutaridine reductase that the physiological forward reaction, but also the reverse reaction from salutaridinol to salutaridine, was catalyzed by the enzyme in the presence of $NADP^+$ and at higher pH (GERARDY & ZENK 1993a, ZIEGLER et al. 2006). The reverse reaction could also possibly occur *in vivo* in animals as a metabolic response to the i.p. injection of [7D]-salutaridinol and would then lead to the loss of deuterium-label forming unlabeled salutaridine that could then be subsequently *N*-demethylated or converted to salutaridinol or thebaine.

The question was whether [7D]-thebaine found in urine of mice i.p. injected with [7D]-salutaridinol was chemically synthesized and would therefore also be detectable after i.p. injection with the heavy-isotope labeled epimer. To address this problem [7D]-*epi*-salutaridinol was synthesized from salutaridine trough reduction with sodium borodeuteride (II.5.3). The application experiment with [7D]-*epi*-salutaridinol was repeated exactly as for [7D]-salutaridinol. A total volume of 11.5 ml urine was collected and subjected to SPE under

a) thebaine standard:

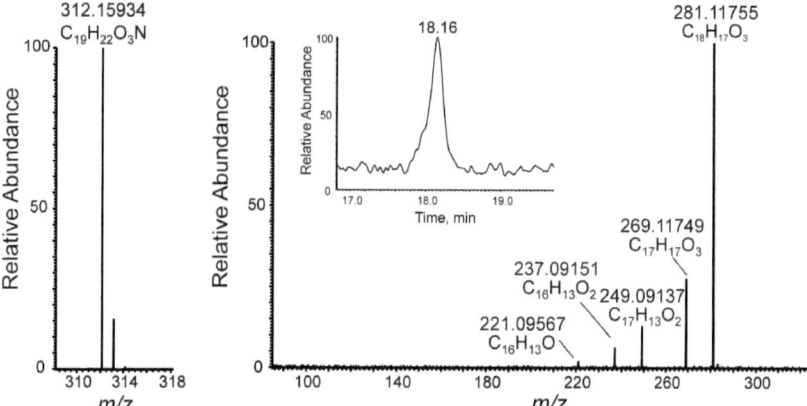

b) [7D]-thebaine isolated from urine of mice i.p. injected with [7D]-salutaridinol:

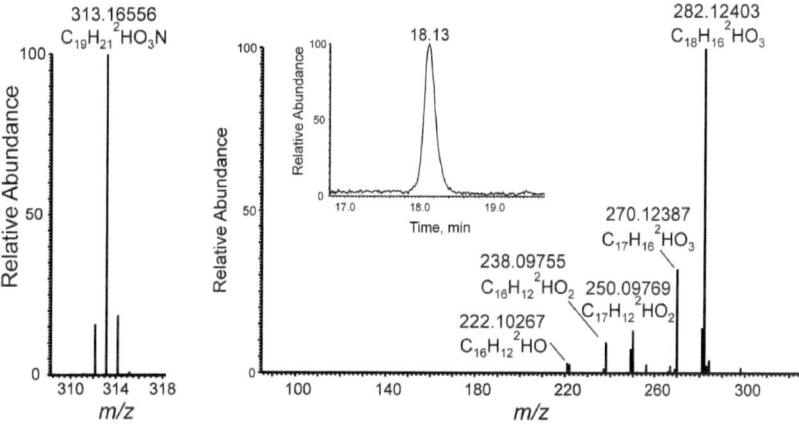

Fig. 18: HR-MS characteristics of a thebaine standard and [7D]-thebaine that was detected in the urine of mice i.p. injected with [7D]-salutaridinol (50 nmol/24 h for 4 days).

a) HR-MS product ion (m/z 312.15934, $C_{19}H_{22}O_3N$), full scan extracted chromatogram (retention time 18.16 min) and MS/MS of thebaine standard.
b) HR-MS product ion (m/z 313.16556, $C_{19}H_{21}{}^2HO_3N$), full scan extracted chromatogram (retention time 18.13 min) and MS/MS of urinary excreted [7D]-thebaine.

Tab. 9: HR-MS data for all metabolites detected in the urine of mice i.p. injected with [7D]-salutaridinol (50 nmol/24 h for 4 days).

Metabolite	Mass measured (m/z)	Mass theoretical (m/z)	Mass accuracy (ppm)	Concentration (per ml urine)
thebaine	312.15939	312.15942	-0.09060	23 pmol (7 ng)
[7D]-thebaine	313.16556	313.16570	-0.43759	128 pmol (40 ng)
demethylated salutaridine	314.13864	314.13869	-0.15734	319 pmol (100 ng)
salutaridine	328.15428	328.15433	-0.17613	275 pmol (90 ng)
salutaridinol	330.16990	330.16999	-0.26848	274 pmol (90 ng)
[7D]-salutaridinol	331.17605	331.17626	-0.647210	2.4 nmol (800 ng)

the same conditions as described for the epimer. HR-LC-MS analysis (II.10.5) of the urine samples revealed that only 0.3% of i.p. injected substance was recovered and was identified as unmetabolized [7D]-*epi*-salutaridinol (97 pmol/ml urine or 32 ng/ml urine). Since no [7D]-thebaine was found to be excreted after administration of [7D]-*epi*-salutaridinol a chemical synthesis of [7D]-thebaine from [7D]-salutaridinol was definitely excluded.

Based on the results described in III.B.2 the *in vivo* formation of thebaine from salutaridine *via* salutaridinol in the living animal is unequivocally confirmed. However, the mechanism behind the formation of the pentacyclic morphine skeleton in animals remains unsolved. The incorporation of the heavy-isotope and the *in vivo* formation of thebaine from salutaridinol but not *epi*-salutaridinol gave proof for the first time that the reaction from salutaridinol to thebaine is enzymatically catalyzed in animals.

3 Morphine formation from Thebaine in Mammals

In opium poppy the terminal steps of morphine biosynthesis start with the entrance of thebaine into the bifurcate pathway (Fig. 1). The first of the two co-existing routes begins with a hydrolysis of the enol ether in thebaine yielding neopinone that is then non-enzymatically isomerized to codeinone with an equilibrium concentration of 42% to 58% (GOLLWITZER *et al.* 1993). Codeinone is then stereoselectively reduced to codeine by the action of codeinone reductase (LENZ & ZENK 1995a) following 6-*O*-demethylation to morphine. In the second alternative pathway first described by BROCHMANN-HANSSEN (1984, 1985) the phenolic ether in thebaine is 6-*O*-demethylated yielding oripavine following a cleavage of the enolic ether of oripavine to morphinone that is then enzymatically reduced to morphine.

The ability of humans to 3-*O*-demethylate codeine to morphine *in vivo* was discovered early in 1954 by MANNERING *et al.* showing that orally administered codeine to human subjects led to the urinary excretion of morphine and, in 1955, by ADLER *et al.* observing that ^{14}C-labeled morphine is excreted by humans after ingestion of ^{14}C-labeled codeine. Interestingly, by monitoring *via* radioimmunoassay with a selective antibody against morphine, a biotranformation of i.p. administered codeine to morphine was observed in rat brain (GINTZLER *et al.* 1976b). The enzyme responsible for the codeine-3-*O*-demethylation in humans was identified as CYP 2D6 (DAYER *et al.* 1988, CHEN *et al.* 1988). This discovery received a lot of attention, especially in clinical aspects of drug therapy, since the analgetic effect of codeine was shown to be dependent on this P450 enzyme (JAFFE & MARTIN 1990) that has been shown to be under genetic control due to a polymorphism (EICHELBAUM & GROSS 1990). CYP 2D6 was also predicted to be active towards the 3-*O*-demethylation of thebaine yielding oripavine (MIKUS *et al.* 1991a). Both 3-*O*-demethylation reactions in the bifurcate pathway were reported as enzymatic activities of porcine salutaridine synthase (AMANN 1994).

With the isolation of morphine from animal tissue such as toad skin (OKA *et al.* 1985) and bovine adrenals and brain (GOLDSTEIN *et al.* 1985) an endogenous formation of mammalian alkaloids *via* a biosynthesis similar to the one in opium poppy was postulated, however, experimental evidence was missing. Because it had been shown that thebaine and codeine are direct precursors in plant morphine biosynthesis both alkaloids, that were reported to be present in mammalian brain (KODAIRA *et al.* 1989), were administered intravenously to rats and, indeed, increased levels of codeine and morphine were observed in rat tissue and body fluids (DONNERER *et al.* 1986). Additionally, *in vitro* experiments with microsomal protein from rat liver, kidney and brain revealed that codeine, oripavine and morphine are formed from thebaine in the presence of an NADPH-generating reaction suggesting that the bifurcate pathway also exists in mammals and that the alkaloids that were detected *in vivo* are formed enzymatically (KODAIRA & SPECTOR 1988). However, none of the labile ketones neopinone, codeinone or morphinone were mentioned in these studies. The isolation of codeinone and morphinone from bile was described later after administration of codeine and morphine to guinea pigs (KUMAGAI et al. 1990, ISHIDA et al. 1991). Moreover, it was shown that the liver of various mammals including humans have the ability to produce morphinone from morphine (YAMANO et al. 1997, TODAKA et al. 2005). This activity was attributed to morphine-6-dehydrogenase, an enzyme that was purified earlier from guinea pig and was shown to oxidize morphine to morphinone and codeine to codeinone at high pH utilizing $NADP^+$ and NAD^+ as cofactors (YAMANO et al. 1985). The enzyme's characteristic to catalyze both forward and reverse reaction, dependent on the pH, was demonstrated with the orthologue purified from hamster liver that reduced morphinone to morphine at pH 7.4 with NADH as cofactor (TODAKA et al. 2000).

Within this thesis the *in vivo* transformation of thebaine to morphine in a living animal and the existence of the bifurcate pathway in mammals as described for human neuroblastoma cells (BOETTCHER et al. 2005) were investigated. For that, the morphine precursors thebaine and oripavine were i.p. injected into mice.

3.1 Intraperitoneal Injection of [*N*-CD$_3$]-Thebaine into Mice

It had been demonstrated by experiments on the pharmacokinetics and metabolism of subcutaneously administered [^{3}H]-thebaine to rats that animals are equipped with the enzymatic system to form the pentacyclic thebaine skeleton (MISRA et al. 1974). These studies revealed that among other metabolites radioactive-labeled oripavine, codeine and morphine were found in rat tissue and urine prompting the postulation of a metabolic pathway of thebaine in animals. The formation of thebaine was shown in this thesis with a sequence of i.p. injections of benzylisoquinoline and morphinan precursors into mice. The next ultimate step was to analyze the metabolites excreted in urine after i.p. injection of thebaine to possibly complete the entire pathway to morphine in mammals starting with norlaudanosoline. Since it was anticipated to verify an endogenous formation of morphine from thebaine in the living animal, a deuterated thebaine standard, that was available in our departmental collection, was administered to mice and urine was analyzed for heavy-isotope labeled morphine metabolites.

For the application experiments, 50 nmol (15.7 µg) [*N*-CD$_3$]-thebaine were i.p. injected into each of four mice every 24 h over 4 days resulting in a total amount of 800 nmol (0.25 mg) salutaridine (each mouse received 2 µmol/kg or 0.63 mg/kg each day). Injected mice were put into a metabolic cage and urine was collected every 24 h and stored at -20°C. A total amount of 16 ml urine was collected over 4 days from the four i.p. injected mice. Morphine and codeine were shown to be excreted by animals in conjugated form as sulfate or glucuronic acid esters (OBERST 1941, SEIBERT et al. 1954, AXELROD & INSCOE 1960, YEH et al. 1977). Since acid hydrolysis was found to be more efficient than enzyme hydrolysis for the cleavage of the conjugates (FISH & HAYES 1974), urine samples obtained from mice i.p. injected with [N-CD$_3$]-thebaine were hydrolyzed in HCl with a final concentration of 2 N at 110°C for 40 min. SPE with Bond Elut Certify cartridges (Varian) was conducted after adjusting the pH to pH 7-8. The preconditioned cartridge was first rinsed and finally eluted under basic conditions according to II.9.3. HR-LC-MS analyses of urine samples were done under the conditions as described in II.10.5. Three urinary metabolites excreted by mice i.p. injected with [N-CD$_3$]-thebaine were identified after comparison of retention time and HR-MS with standards as [N-CD$_3$]-codeine (Fig. 19a,b), [N-CD$_3$]-oripavine (Fig. 19c,d) and [N-CD$_3$]-morphine (Fig. 19e,f). All three products were verified by HR-MS with a high mass accuracy of < 1 ppm (Tab. 11). However, the quantitative analysis revealed that only very low concentrations of these products (below 3 pmol/ml urine or 1 ng/ml urine) were isolated from urine of the injected mice (Tab. 10) resulting in a total of only 0.01% of i.p. injected substance that could be recovered. A possibly yet undiscovered metabolic fate of thebaine in animals and degradation of excreted thebaine maybe due to acid hydrolysis as described by KODAIRA et al. (1989) could account for levels below the limit of detection.

Among all the application studies conducted in this thesis this is the first time that morphine was found to be excreted in the urine of i.p. injected mice. Moreover, due to the incorporation of three deuterium atoms it was also shown that the living animal is capable of an endogenous formation of morphine from thebaine. The hypothesis that the biosynthesis of morphine in animals occurs similarly to the one in opium poppy *via* two alternative pathways (KODAIRA & SPECTOR 1988) can now be supported due to the findings of the intermediates [N-CD$_3$]-oripavine and [N-CD$_3$]-codeine following i.p. injection of [N-CD$_3$]-thebaine.

3.2 Intraperitoneal Injection of Oripavine into Mice

The formation of morphine from codeine in mammals has been described extensively especially after the enzyme that catalyzes this conversion, CYP 2D6, had been found. However, there is a lack of experiments describing the second alternative pathway that produces morphine through the 6-*O*-demethylation of oripavine. To support the claim that this alternative pathway not only exists in plants but also in animals, application experiments with highly purified, morphine-free oripavine as determined by HR-MS (II.10.5) were conducted.

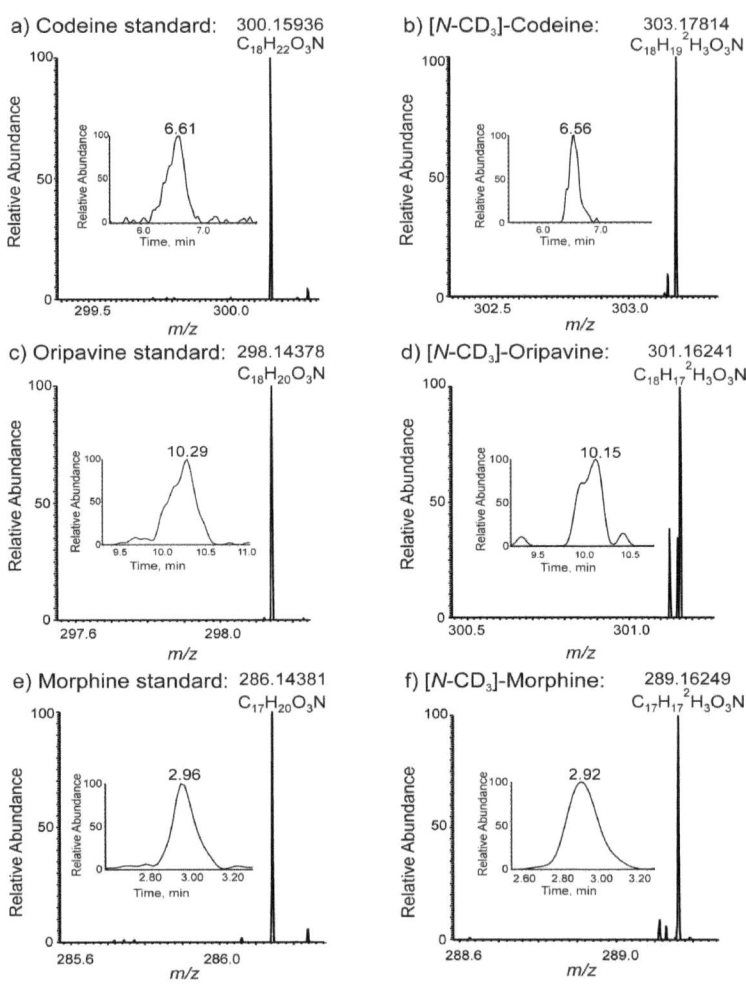

Fig. 19: HR-MS characteristics of standards and heavy-isotope labeled metabolites detected in the urine of mice i.p. injected with [N-CD$_3$]-thebaine (50 nmol/24 h for 4 days).

a) HR-MS product ion and full scan extracted chromatogram of codeine standard.
b) HR-MS product ion and full scan extracted chromatogram of [N-CD$_3$]-codeine isolated from urine of mice i.p. injected with [N-CD$_3$]-thebaine.
c) HR-MS product ion and full scan extracted chromatogram of oripavine standard.
d) HR-MS product ion and full scan extracted chromatogram of [N-CD$_3$]-oripavine isolated from urine of mice i.p. injected with [N-CD$_3$]-thebaine.
e) HR-MS product ion and full scan extracted chromatogram of morphine standard.
f) HR-MS product ion and full scan extracted chromatogram of [N-CD$_3$]-morphine isolated from urine of mice i.p. injected with [N-CD$_3$]-thebaine.

Tab. 10: HR-MS data for [N-CD$_3$]-codeine, [N-CD$_3$]-oripavine and [N-CD$_3$]-morphine detected in the urine of mice i.p. injected with [N-CD$_3$]-thebaine (50 nmol/24 h for 4 days).

Metabolite	Mass measured (m/z)	Mass theoretical (m/z)	Mass accuracy (ppm)	Concentration (per ml urine)
[N-CD$_3$]-codeine	303.17814	303.17825	-0.31991	0.7 pmol (0.2 ng)
[N-CD$_3$]-oripavine	301.16241	301.16260	-0.69736	2.2 pmol (0.7 ng)
[N-CD$_3$]-morphine	289.16249	289.16260	-0.38912	1.7 pmol (0.49 ng)

I.p. injection of 50 nmol (14.9 µg) oripavine into two mice every 24 h over 4 days resulted in a total amount of 400 nmol (0.12 mg) oripavine (each mouse received 2 µmol/kg or 0.6 mg/kg each day). Injected mice were put into a metabolic cage and urine was collected every 24 h and stored at -20°C. A total amount of 6.8 ml urine was collected over 4 days from the two i.p. injected mice. Urine samples were worked-up by starting with a hydrolysis in HCl with a final concentration of 2 N at 110°C for 40 min. A SPE with Bond Elut Certify cartridges (Varian) followed after adjusting the pH to 7-8. The preconditioned cartridge was first rinsed and finally eluted under basic conditions (II.9.3). HR-LC-MS analyses of urine samples were done under the conditions as described in II.10.5.

Three metabolites were detected in urine of mice i.p. injected with oripavine that did not occur in control urine. The quantitative analysis revealed that only 0.3% of the total amount of i.p. injected substance was recovered. One of the three urinary products was unequivocally verified after comparison of retention time and HR-MS with a standard as morphine (-0.31073 ppm) in a concentration of 1.1 pmol/ ml urine (0.3 ng/ml urine) (Fig. 20a). The biotranformation of oripavine to morphine as demonstrated here confirms for the first time that animals are capable of catalyzing the cleavage of the vinyl-ether in oripavine.

Next to morphine, a second product with the same mass as morphine but different retention time was detected but could not be clearly assigned. A third urinary metabolite with m/z 284.12803 (-0.31188 ppm) that was found to be present after i.p. injection of oripavine was predicted to be morphinone (Fig. 20b). This assumption is supported by reports describing morphinone as an *in vivo* and *in vitro* metabolite of morphine in mammals (YAMANO *et al.* 1997, TODAKA *et al.* 2005) but also as the substrate for the reduction reaction

a) Morphine isolated from urine of mice i.p. injected with Oripavine:

b) Morphinone isolated from urine of mice i.p. injected with Oripavine:

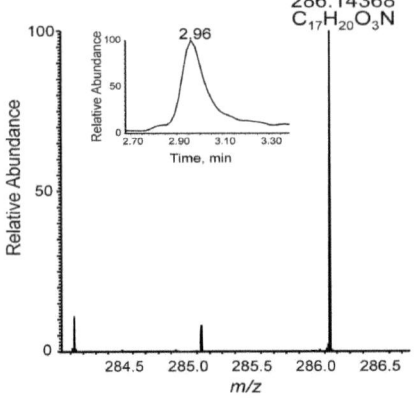
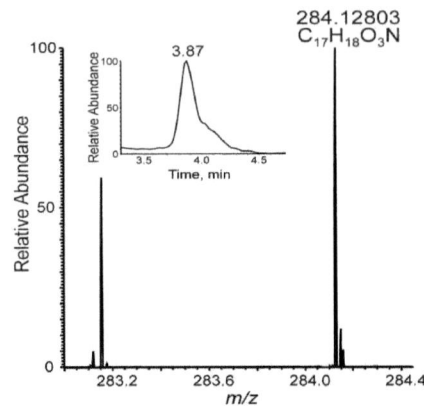

Fig. 20: HR-MS product ion and full scan extracted chromatogram of two metabolites found in urine of mice i.p. injected with oripavine (50 nmol/24 h for 4 days).

a) Morphine was isolated as a urinary metabolite from the i.p. injection of oripavine and verified by comparison with a morphine standard (see also Fig. 19e)
b) Morphinone was strongly suggested to be a urinary product after i.p. injection of oripavine into mice. However, a standard was not available to confirm this prediction.

to morphine (TODAKA et al. 2000) indicating that there is an enzyme system present in mammals which catalyzes the final step of morphine biosynthesis involving the 6-O-demethylation of oripavine.

With the i.p. injection of simple benzylisoquinolines and morphinan alkaloids to mice described in Part B of this thesis it was shown for the first time that the living animal is capable of biosynthesizing morphine. It was shown that (R)-norlaudanosoline is the most plausible natural precursor for the morphinan alkaloid salutaridine. Application of (R)-[N-CD$_3$]-reticuline led to the urinary excretion of a set of phenol-coupled products, including salutaridine, that all carried the heavy-isotope label. The capability of living animals to catalyze the closure of the oxide bridge to produce thebaine from salutaridine and salutaridinol, a biochemically challenging reaction, was clearly demonstrated with the i.p. injection of salutaridine and [7D]-salutaridinol. Finally, the existence of a bifurcate pathway to form morphine in animals was confirmed by application experiments with [N-CD$_3$]-thebaine and oripavine.

Part C: Enzyme Studies in Selected Morphine Biosynthetic Steps

In vitro studies as they are described in result part B of this thesis are an essential tool to reveal the biosynthesis of morphine in mammals considering that feeding of morphine precursors to opium poppy elucidated the pathway in plants. In fact, feeding experiments with human neuroblastoma cells conducted by the group around M.H. ZENK (POEAKNAPO *et al.* 2004, BOETTCHER *et al.* 2005) demonstrated that these human cells are equipped with the complete enzymatic systems to produce morphine from tyrosine. The biochemical approach described within this thesis of i.p. injecting potential precursor solutions into mice combined to an efficient extraction procedure and to the outstanding analytical method of HR-MS showed unequivocally and for the first time the *in vivo* formation of morphine in the living animal from the distant precursor and central intermediate norlaudanosoline. However, the involved mammalian catalysts and their reaction mechanisms are only poorly understood. The complex knowledge about biosynthetic routes towards morphine in animals can be accomplished and verified by combining the results from the *in vivo* experiments with enzymatic studies of selected biosynthetic steps.

The finding that i.p. injections of norlaudanosoline resulted in the urinary excretion of salutaridine (III.B.1) prompted a search for the mammalian enzymes that are involved in the formation of norlaudanosoline and in the three methylation steps that are required to form reticuline, the substrate for the phenol-coupling reaction yielding salutaridine. For that, norlaudanosoline needs to be *O*-methylated at the hydroxy group on C4′ and C6. The enzyme that is known to catalyze the 6-*O*-methylation of benzylisoquinolines is catechol-*O*-methyltransferase (COLLINS *et al.* 1973). A closer study of the regioselectivity of the *O*-methylation of norlaudanosoline revealed that *in vivo* in rat brain predominantly the hydroxy groups at position C3′,C6 and C7 of the THBIQ skeleton are monomethylated and to a minimal extent also the hydroxy group at C4′ (CASHAW *et al.* 1983). The enzyme that catalyzes the 4′-*O*-methylation is so far not known. However, a 4′-*O*-methylation activity of THBIQ alkaloids was not only observed by CASHAW *et al.* (1983) but also in this thesis indirectly with the urinary excretion of salutaridine, scoulerine and coreximine after i.p. injection of norlaudanosoline into mice. These three alkaloids carry a methoxy group that can only originate from an earlier 4′-*O*-methylation. The third methylation that is required to build reticuline from norlaudanosoline is the methylation of the ring nitrogen in the isoquinoline moiety of the THBIQ molecule which was closer investigated within this thesis. Evidence for enzymatically catalyzed reactions in the terminal steps of morphine biosynthesis in animals was reported by research of the two groups around A. GOLDSTEIN and S. SPECTOR (WEITZ et al. 1987, KODAIRA & SPECTOR 1988). The first enzyme purified from pig liver catalyzing the key reaction in morphine biosynthesis was discovered by the group of M.H. ZENK (AMANN et al. 1995). The P450 enzyme was not only capable of catalyzing the phenol-coupling reaction of (R)-reticuline to salutaridine but also the 3-O-demethylation of thebaine to oripavine and codeine to morphine in the bifurcate pathway. A homology search revealed high sequence identity to CYP 2D6, a human P450 enzyme that was already known for catalyzing the 3-O-demethylation of thebaine to oripavine (MIKUS et al. 1991a) and codeine to morphine (DAYER et al. 1988, CHEN et al. 1988). In the following chapter, mammalian enzymes catalyzing salutaridine synthesis from (R)-reticuline were further investigated.

This part of the thesis attempted the identification of new enzymes possibly involved in the biosynthesis of morphine in animals as well as the characterization of their properties and reaction mechanisms. The sequence of the enzymatic studies described next corresponds to the sequence of reactions as they are postulated to occur in the pathway of morphine in mammals.

1 Formation of 3,4-Dihydroxyphenylacetaldehyde (DOPAL) in Animals

It was reported by HOLTZ et al. (1964) and DAVIS & WALSH (1970) that the alkaloid norlaudanosoline, the first intermediate in mammalian morphine biosynthesis, is easily formed non-enzymatically in animals by the condensation of 3,4-dihydroxyphenylacetaldehyde (DOPAL) with dopamine. Both in vitro studies showed DOPAL as a metabolite of dopamine catabolism in animals and identified the aldehyde as the product of the reaction of dopamine via monoamine oxidase. Two reactive groups, a catechol and an aldehyde group, that can interact with sulfhydryl- and amino groups of biomolecules such as proteins and nucleic acids, make the aldehyde a very unstable and highly reactive molecule. In animal tissue, DOPAL is mainly oxidized to 3,4-dihydroxyphenylacetic acid via the action of NAD-dependent aldehyde dehydrogenase (E.C. 1.2.1.3) or to a lesser extent enzymatically reduced to 3,4-dihydroxyphenylethanol (DOPET) by aldehyde reductase (EC 1.1.1.2) (BREESE et al. 1969, TABAKOFF et al. 1973). Another pathway for the formation of DOPAL other than the one suggested previously from dopamine was investigated in this thesis. Instead of assuming that DOPAL in morphine biosynthesis is provided by oxidative deamination of dopamine the possibility of a new route was further explored, starting from L-DOPA that

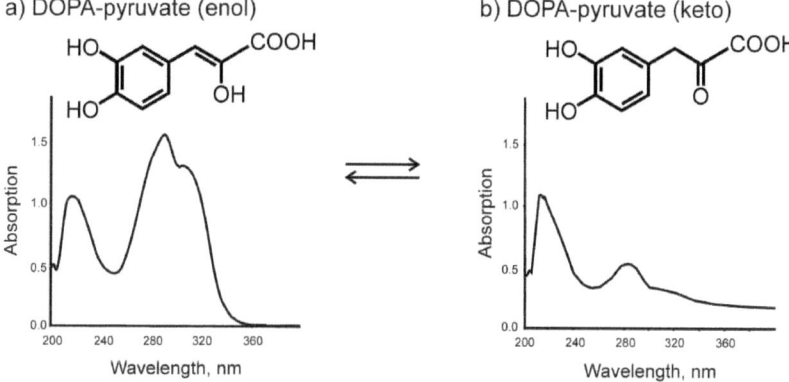

Fig. 21: Spectrophotometric detection of the tautomerization of DOPA-pyruvate. While the enol absorbs intensively in the ultraviolet region the keto form does not.
assay: 18 µg DOPA-pyruvate in 1 ml 0.05 M acetic acid pH 6.0
incubation: 2 h, room temperature

a) Enol form of DOPA-pyruvate at start of incubation.
b) Keto form of DOPA-pyruvate 2 h after incubation.

through transamination yields 3,4-dihydroxyphenylpyruvate (DOPA-pyruvate) which is then decarboxylated to the aldehyde. For the decarboxylation reaction, DOPA-pyruvate needs to be completely tautomerized into the keto form. To allow the formation of the required keto form, DOPA-pyruvate was prepared freshly before the reaction in 0.05 M acetic acid pH 6 as it was described for *para*-hydroxyphenylpyruvate by KNOX & EDWARDS (1955). The conversion of the enol form into the keto form can be followed spectrophotometrically by a decrease of absorption at 290-310 nm (Fig. 21). Whereas *para*-hydroxyphenylpyruvate was suggested to be left at room temperature for 24 h to tautomerize (KNOX & EDWARDS 1955) DOPA-pyruvate was present in its keto-form after 2 h.

The 4000 QTRAP MS instrument was used as analytical detection method to increase sensitivity and to exclude any interference in the detection of products formed from DOPA-pyruvate other than DOPAL. The enhanced product ion (EPI) analysis in negative ionization mode obtained with the 4000 QTRAP MS instrument showed for DOPAL (m/z 151) two fragment ions m/z 123 and m/z 122 (Fig. 22a). The two transitions of m/z 151 to m/z 123 and m/z 151 to m/z 122 were chosen for the multiple reaction monitoring (MRM) experiment to identify DOPAL in a reaction mixture (Fig. 22b).

Cytosolic mouse liver protein was prepared according to II.7.2 and tested with DOPA-pyruvate (keto form) as substrate as follows:

assay:

50 mM sodium pyrophosphate pH 8.8	433 µl	(21.65 mM)
5 mM DOPA-pyruvate (keto)	33 µl	(165 nmol)
0.1 M ThDP	10 µl	(1 µmol)
0.1 M MgCl$_2$	10 µl	(1 µmol)
16 mg/ml soluble mouse liver protein	62 µl	(0.5 mg)
ad dd H$_2$O	1 ml	

control assay: with boiled protein
incubation: 37°C, 240 min
detection: QTRAP 4000 LC-MS with Eclipse HPLC column (II.10.4.A)

After the incubation, enzyme assays were terminated by addition of 100 µl 10% TCA. The reaction mixture was vortexed, centrifuged for 3 min at 13200 rpm and 10 µl of the supernatant was subjected to LC-MS/MS analysis under the conditions as described in detail in II.10.4.A. DOPAL was unequivocally identified as a product from DOPA-pyruvate (Fig. 22c) and absent in control assays with boiled protein (Fig. 22d).

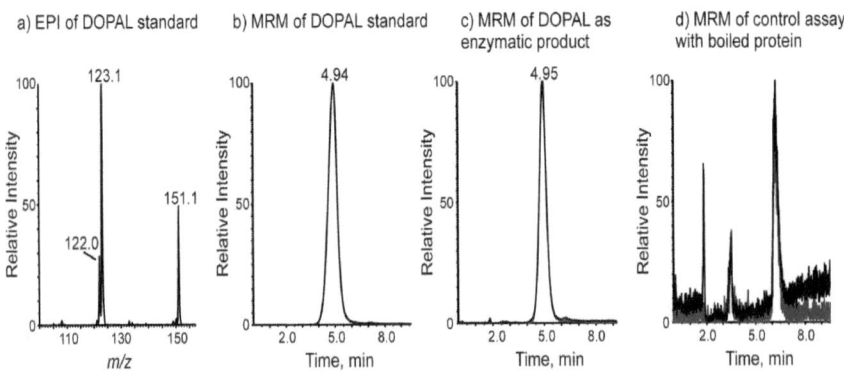

Fig. 22: The LC-MS analysis of DOPAL reveals that the aldehyde is produced from DOPA-pyruvate.
assay: 21.65 mM sodium pyrophosphate pH 8.8, 165 nmol DOPA-pyruvate (keto), 1 µmol ThDP, 1 µmol MgCl$_2$ and 1 mg soluble mouse liver protein
incubation: 2 h, 37°C

a) EPI (collision energy -30 V, declustering potential -70 V) of DOPAL standard [M-H]⁻ 151 showed two main fragments m/z 122 and m/z 123.
b) MRM of DOPAL standard with the two characteristic transitions of m/z 151 to m/z 123 and m/z 151 to m/z 122 that both superimposed.
c) MRM of enzymatically produced DOPAL.
d) MRM of assays with boiled mouse liver protein.

In addition, the reaction mixture, that was previously shown to produce DOPAL from DOPA-pyruvate in presence of mouse liver protein (Fig. 22c), was incubated again in the presence of 25.5 units horse liver alcohol dehydrogenase (ADH, Sigma) and 200 nmol NADH. A formation of DOPET via reduction of the aldehyde catalyzed by ADH was expected to occur in this second incubation experiment. To follow the product formation of DOPET by LC-MS with a 4000 QTRAP MS instrument, a standard was first chemically synthesized by reduction of 200 µg DOPAL with a small amount of sodium borohydride in 70% methanol. Subsequently, EPI and MRM were determined for the synthesized alcohol (m/z 153) in negative ionization mode. As shown in Fig. 23a, DOPET also fragmented like the aldehyde to two ions with m/z 123 and m/z 122 that were chosen as transitions for the MRM experiment (Fig. 23b). A formation of DOPET from DOPA-pyruvate due to the reduction of the aldehyde by ADH was indeed observed in incubation mixture containing ADH and NADH (Fig. 23c). The alcohol was absent in control assays with boiled mouse liver protein (Fig. 23d). When ADH and NADH were excluded from the incubation mixture low amounts of DOPET were also detectable probably catalyzed by endogenous mouse liver aldehyde reductase.

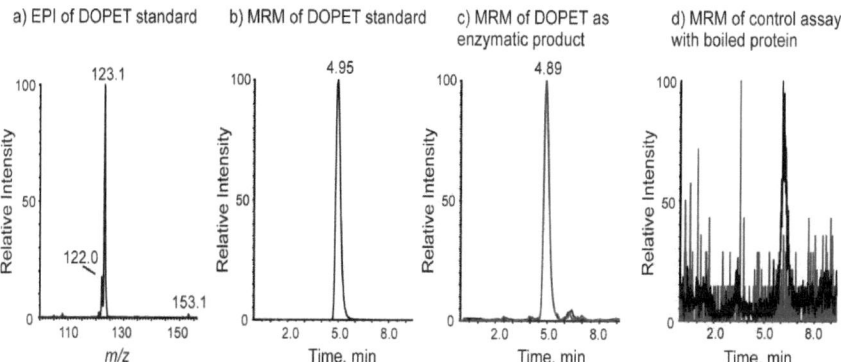

Fig. 23: LC-MS analysis of DOPET formed from DOPAL catalyzed by ADH.
assay: 21.65 mM sodium pyrophosphate pH 8.8, 165 nmol DOPA-pyruvate (keto), 1 µmol ThDP, 1 µmol MgCl$_2$, 200 nmol NADH, 25.5 units ADH and 1 mg soluble mouse liver protein
incubation: 2 h, 37°C

a) EPI (collision energy -30 V, declustering potential -70 V) of DOPET standard ([M-H]$^-$ 153) revealed two main fragment ions m/z 122 and m/z 123.
b) MRM of DOPET standard with the two compound specific transitions m/z 153 to m/z 123 and m/z 153 to m/z 122 that both superimposed.
c) MRM of enzymatically formed DOPET.
d) MRM of reaction mixtures containing boiled mouse liver protein.

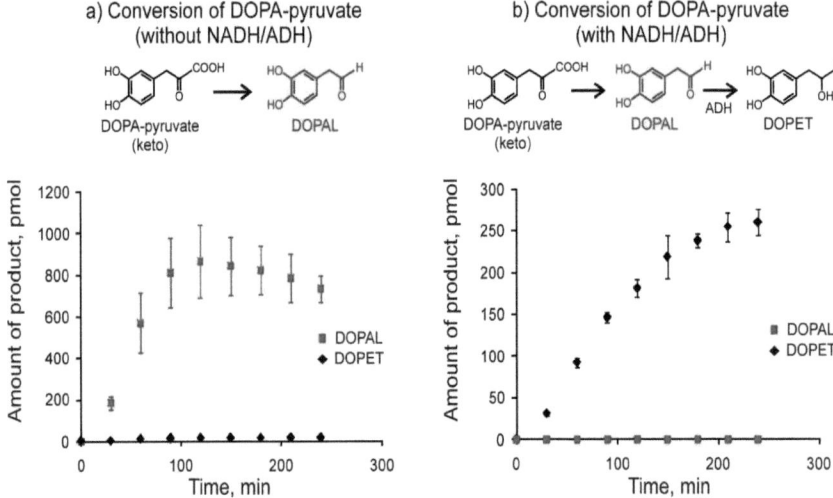

Fig. 24: Time-dependent conversion of DOPA-pyruvate to DOPAL and DOPET.
assay: 165 nmol DOPA-pyruvate (keto), 21.65 mM sodium pyrophosphate pH 8.8, 1 µmol ThDP, 1 µmol $MgCl_2$ and 1 mg soluble mouse liver protein.
incubation: 0-240 min, 37°C

a) Without ADH/NADH.
b) With 200 nmol NADH and 25.5 units ADH.

Time curve assays were recorded for the formation of DOPAL and DOPET with soluble mouse liver protein and without or with ADH (Fig. 24). It was clearly shown that DOPAL and low amounts of DOPET are produced in the absence of ADH and NADH (Fig. 24a). However, addition of ADH and NADH to the reaction system led to an increase of the formation of DOPET and, ultimately, to the absence of DOPAL due to its complete conversion to the alcohol catalzyed by ADH (Fig. 24b). The studies described herein showed that animals are capable of catalyzing the decarboxylation of DOPA-pyruvate to DOPAL as an alternative route for the production of the aldehyde.

2 Condensation to Norlaudanosoline

A special case of the Mannich reaction is the Pictet-Spengler reaction, the condensation of β-arylethylamines with carbonyl compounds, named after the two chemists Amé Pictet and Theodor Spengler that discovered the reaction in 1911. In the case of the entry reaction in plant morphine biosynthesis (*S*)-norcoclaurine is produced by condensation of dopamine as β-arylethylamine with the carbonyl compound *para*-hydroxyphenylacetaldehyde (STADLER *et al.* 1989). The stereoselective Pictet-Spengler type reaction was found to be catalyzed by two completely unrelated plant enzymes, one pathogenesis related enzyme isolated from *Talictrum flavum* suspension cultures and the other enzyme from *Coptis japonica* with a sequence similarity to 2-oxoglutarate-dependent dioxygenases (SAMANANI *et al.* 2004, MINAMI *et al.* 2007).

Among the reports on the occurrence of norlaudanosoline in animals (Tab. 6) two researchers described the detection of the (S)-enantiomer in brain of ethanol treated rats (HABER *et al.* 1997) or naturally occurring in human brain (SANGO *et al.* 2000). Additionally, BOETTCHER *et al.* (2005) predicted the (S)-enantiomer to be the first alkaloid intermediate in human morphine biosynthesis. The question, therefore, whether the condensation reaction in animals is similar to plants catalyzed by a stereospecific enzyme was attempted to be answered in this thesis by chiral separation of the reaction product norlaudanosoline. Since a chemical reaction leads to equal amounts of (R)- and (S)-norlaudanosoline it was analyzed if addition of mouse liver protein leads to a shift of the racemic equilibrium caused by a possible stereoselective enzyme catalyzing the condensation reaction. For that, mouse liver protein was prepared according to II.7.1 and incubated in following reaction mixture:

assay: 1 M potassium phosphate pH 7.4   10 µl   (100 mM)
1 mM DOPAL   6.6 µl   (6.6 nmol)
1 mM Dopamine   6.6 µl   (6.6 nmol)
83 mg/ml mouse liver protein   0.1-1 mg
ad dd H$_2$O   100 µl

control assay: with boiled or absent protein
incubation: 37°C, 60 min
detection: QTRAP 4000 LC-MS with Chiral-CBH HPLC column (II.10.4.B)

The incubation reactions were terminated by addition of 10 µl of 20% TCA. After centrifugation, the pH of the supernatant was adjusted to pH 7 with KOH and subjected to a LC-MS/MS with a Chiral-CBH HPLC column from Chromtech (II.10.4.B). The chiral selector of the column is cellobiohydrolase, an immobilized enzyme that works between pH 3 to 7. The column achieves the chiral separation of a multitude of compounds including (R)- and (S)- norlaudanosoline (Fig. 25a). The spontaneous reaction showed indeed an equal amount of both enantiomers (Fig. 25b) that was not changed after addition of 0.1 mg protein (Fig. 25c) or 1 mg protein indicating that the condensation reaction was non-enzymatically catalyzed (spontaneous reaction).

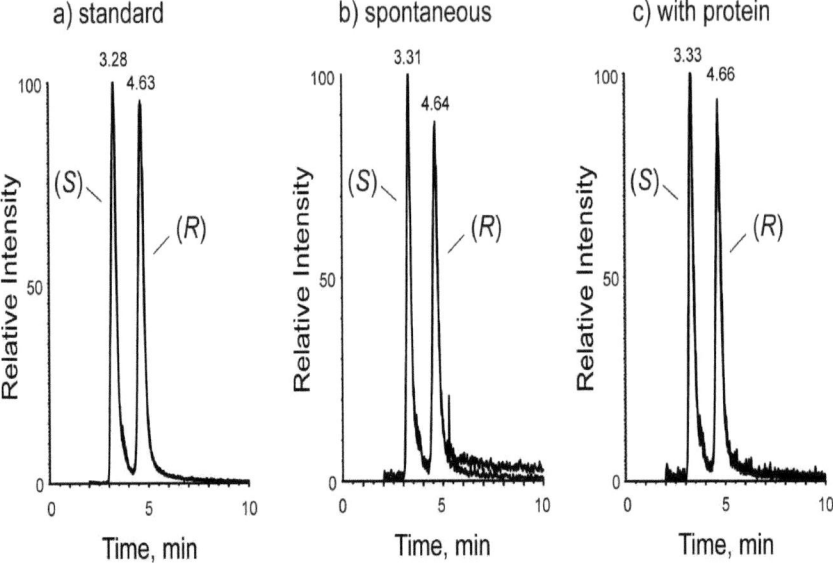

Fig. 25: Chiral separation of (*R,S*)-norlaudanosoline by LC-MS.
assay: 6.6 nmol DOPAL, 6.6 nmol dopamine, 100 mM potassium phosphate pH 7.4
incubation: 60 min, 37°C

a) Standard (*R,S*)-norlaudanosoline (peak area 4.4 x10^5 (*S*)- and 4.3 x 10^5 (*R*)-enantiomer).
b) Spontaneous reaction without protein (peak area 2.1 x10^5 (*S*)- and 2.3 x 10^5 (*R*)-enantiomer).
c) Reaction in the presence of mouse liver protein (peak area 3.8x10^5 (*S*)- and 3.9 x 10^5 (*R*)-enantiomer).

In animals, it had been shown that incubation of kidney homogenates with L-DOPA produced a compound, exhibiting blood pressure lowering properties, which was assigned to norlaudanosoline (HOLTZ & HEISE 1938, HOLTZ et al. 1963). The existence of the route that produces DOPAL from L-DOPA *via* DOPA-pyruvate was further investigated within this part of the thesis by analyzing position-specific incorporation of heavy-isotope labeled L-DOPA into norlaudanosoline. An enzyme assay was conducted in which mouse liver protein was incubated with dopamine and heavy-isotope labeled [ring-$^{13}C_6$]-L-DOPA. According to the fragmentation pathway of norlaudanosoline as well as the heavy-isotope label of the substrate a specific MRM experiment was designed to cover all possible transitions of unlabeled as well as of three differently labeled condensation products (Fig. 26a). The following test assay was prepared:

assay 1: 1 M sodium pyrophosphate pH 7.4 20 µl (20 mM)
10 mM [ring-$^{13}C_6$]-L-DOPA 33 µl (330 nmol)
10 mM dopamine 33 µl (330 nmol)
0.1 M ThDP 10 µl (1 µmol)
0.1 M $MgCl_2$ 10 µl (1 µmol)
45 mg/ml mouse liver protein 100 µl (*ca.* 0.5 mg)
ad dd H_2O 1 ml

control assay: without protein
incubation: 37°C, 120 min
detection: 4000 QTRAP LC-MS with Eclipse HPLC column (II.10.4.A)

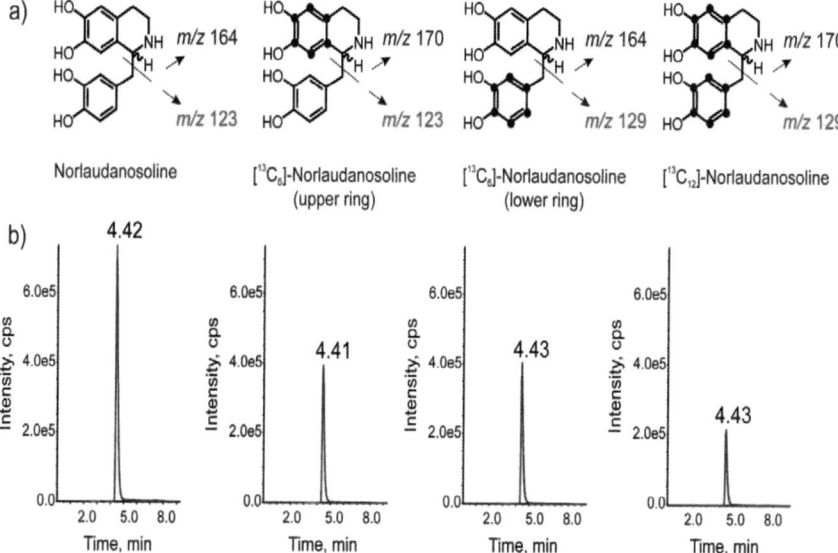

Fig. 26: Formation of norlaudanosoline from [$^{13}C_6$]-L-DOPA incubated together with mouse liver protein and dopamine.
  assay: 20 mM potassium phosphate pH 7.4, 330 nmol [$^{13}C_6$]-L-DOPA, 330 nmol dopamine, 1 µmol ThDP, 1 µmol $MgCl_2$, 0.5 mg mouse liver protein.
  incubation: 120 min, 37°C

a) Labeling and fragmentation of four molecules of norlaudanosoline possibly produced by the reaction of dopamine with [$^{13}C_6$]-L-DOPA.
b) MRM transitions (isoquinoline fragment labeled in blue, benzyl fragment labeled in red, both transitions superimpose) corresponding to the fragments shown in a).

The reaction mixtures were terminated by adding 100 µl 20% TCA. After precipitation and centrifugation the supernatant was subjected to LC-MS/MS with an Eclipse DB-C18 HPLC column under the conditions as described in II.10.4.A. As shown in Fig. 26b unlabeled norlaudanosoline as well as [$^{13}C_6$]- and [$^{13}C_{12}$]-labeled norlaudanosoline were found to be present after incubation of [ring-$^{13}C_6$]-L-DOPA with dopamine. Unlabeled norlaudanosoline was interpreted as a condensation product of dopamine with unlabeled DOPAL formed by monoamine oxidase from unlabeled dopamine. Two products were detected for [$^{13}C_6$]-labeled norlaudanosoline with the label in the upper ring, the isoquinoline moiety, and the second product with the label in the benzyl moiety. The finding of [$^{13}C_6$]-norlaudanosoline with the label in the isoquinoline moiety could be explained by the condensation of [ring-$^{13}C_6$]-dopamine formed from [ring-$^{13}C_6$]-L-DOPA with unlabeled DOPAL produced from unlabeled dopamine *via* the action of monoamine oxidase. The [$^{13}C_6$]-label in the benzyl ring of norlaudanosoline clearly confirmed the existence of the route that forms [ring-$^{13}C_6$]-DOPAL from [ring-$^{13}C_6$]-L-DOPA through transamination and decarboxylation. The formation of [$^{13}C_{12}$]-norlaudanosoline was explained by a condensation of [ring-$^{13}C_6$]-dopamine with a [ring-$^{13}C_6$]-labeled aldehyde deriving either from [ring-$^{13}C_6$]-dopamine through the action of monoamine oxidase or from [ring-$^{13}C_6$]-DOPA-pyruvate by decarboxylation. None of these products were observed in the control assay. Chiral LC-MS/MS analysis of the sample (II.10.4.B) revealed that equal amounts of (*R*)- and (*S*)-norlaudanosoline were produced. This result suggested two possibilities: either that unlike in plants the condensation reaction in animals might not be stereospecific or that the activity of a stereospecific enzyme catalyzing the condensation reaction was too low to be detected in incubation mixtures containing crude mouse liver protein.

3 *N*-Methylation in Mammalian Morphine Biosynthesis

One of the methylation steps required for the formation of reticuline from norlaudanosoline in mammalian morphine biosynthesis is the methylation at the ring nitrogen. Two *N*-methyltransferases were found in the database of the human genome that could be capable to *N*-methylate low molecular weight compounds such as THBIQ alkaloids. Whereas phenylethanolamine *N*-methyltransferase (PNMT) (EC 2.1.1.28) exhibits a relatively strict substrate specificity and represents a pharmacological target by being involved in the final step of the biosynthesis of the neurotransmitter adrenaline (AXELROD 1962) the second enzyme named amine *N*-methyltransferase (NMT) (EC 2.1.1.49) has practically no functional role *in vivo*. Amine *N*-methyltransferase was first described as rabbit lung methyltransferase by AXELROD (1961) and since it had very broad substrate specificity it was predicted to be involved in the metabolism of numerous drugs, xenobiotics and endogenous molecules. The ability to *N*-methylate indolethylamines such as tryptamine raised a special interest since a possible participation in the formation of psychoactive agents as well as neurotoxins was anticipated. However, particularly after the rabbit and human genes were cloned by the group around R.M. WEINSHILBOUM and an apparent K_m of 2.9 mM for tryptamine as the best substrate for the recombinant enzyme was determined this possibility seemed to be less likely (THOMPSON & WEINSHILBOUM 1998, THOMPSON *et al.* 1999). Among all the primary and secondary amines that were tested tetrahydoisoquinoline alkaloids (TIQ) were also found to be substrates of the enzyme (ANSHER & JAKOBY 1986, THOMPSON & WEINSHILBOUM 1998). Because of a special interest in the *N*-methylation of TIQ compounds in animals, that were suggested to be possibly involved in mechanisms of Parkinson's disease, BAHNMAIER *et al.* (1999) purified the enzyme from bovine liver. With the goal to determine the stereospecificity of NMT they tested enantiomers of two methylated TIQ derivatives and found a selective stereospecificity depending on the degree of methylation of the substrate. If the TIQ contained two vicinal methoxy groups like in salsolidine the (*S*)-enantiomer was preferred whereas in the case of only one methoxy group as in isosalsoline the (*R*)-enantiomer was preferred.

3.1 Heterologous Expression of two Human *N*-Methyltransferases

A cDNA clone of PNMT was generously provided as glycerol stock by Dr. X. Ren, Donald Danforth Plant Science Center (St. Louis, USA). A cDNA clone of the other *N*-methyltransferase, NMT, was obtained from the Deutsche Ressourcenzentrum für Genomforschung, Berlin, Germany (RZPD, clone BC 106902.1). The sequence of this clone was verified and shown to contain three amino acids that were different from the cDNA published by THOMPSON *et al.* (1999). The authors explained that single-nucleotide polymorphisms could cause the replacement of single pyrimidine or purine bases and could therefore lead to this DNA variation. A sequence alignment of human PNMT, rabbit NMT, human NMT (THOMPSON *et al.* 1999) and human NMT (RZPD) is shown in Fig. 27. In the methylation reaction the enzyme catalyzes the transfer of a methyl group from the donor *S*-adenosyl-L-methionine (SAM) to the substrate. Three motifs (I, II and III) that are labeled in Fig. 27 with a black bar are suggested to be common to SAM-dependent methyltransferases and to be involved in binding of the co-substrate (KAGAN & CLARKE 1994).

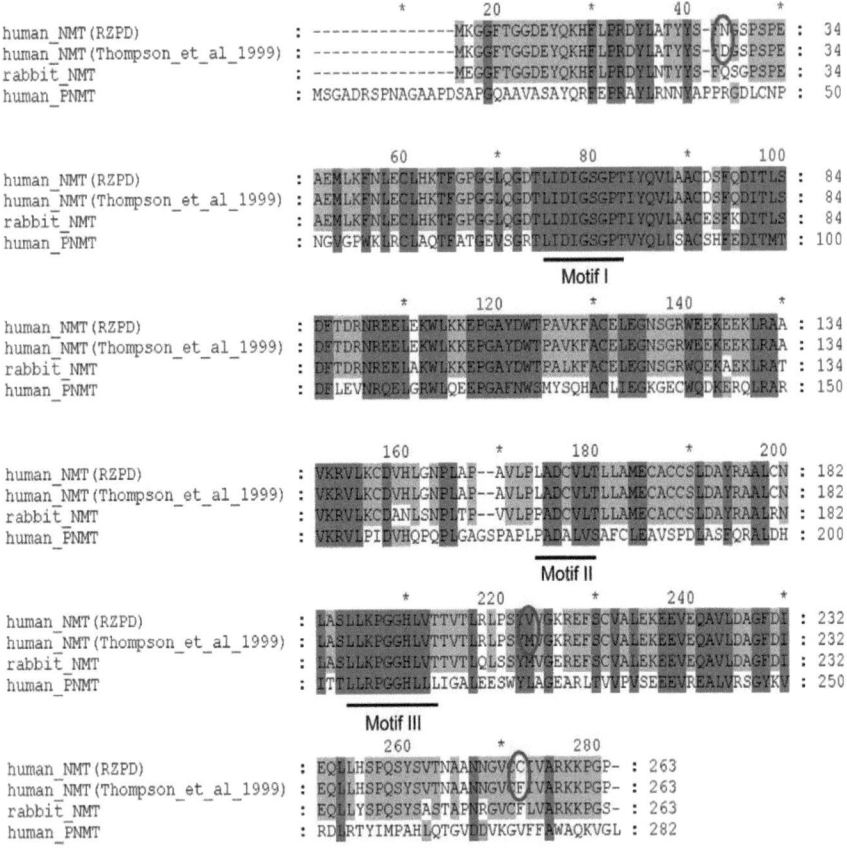

Fig. 27: Sequences of human NMT (RZPD), human NMT (THOMPSON et al. 1999), rabbit NMT and human PNMT aligned by "ClustalW" and edited with "GeneDoc". Marked in dark grey are conserved sequences with an identity of 80%; 60% identical sequences are shown in light grey. Circled in red are three pairs of amino acids that differ between the human NMT clone described in this thesis (obtained from RZPD) and human NMT clone as published by THOMPSON et al. (1999). Motifs I, II and III are conserved sequences that were suggested to be involved in the binding of the co-substrate SAM (KAGAN & CLARKE 1994).

The cDNA was cloned into expression vector pET28a that added a polyhistidine-tag (His-tag). These additional histidine residues make it possible to easily purify the protein *via* metal affinity chromatography. Since human PNMT was provided by Dr. X. Ren in the right vector, only human NMT needed to be cloned which is described in detail in II.7.3.1.

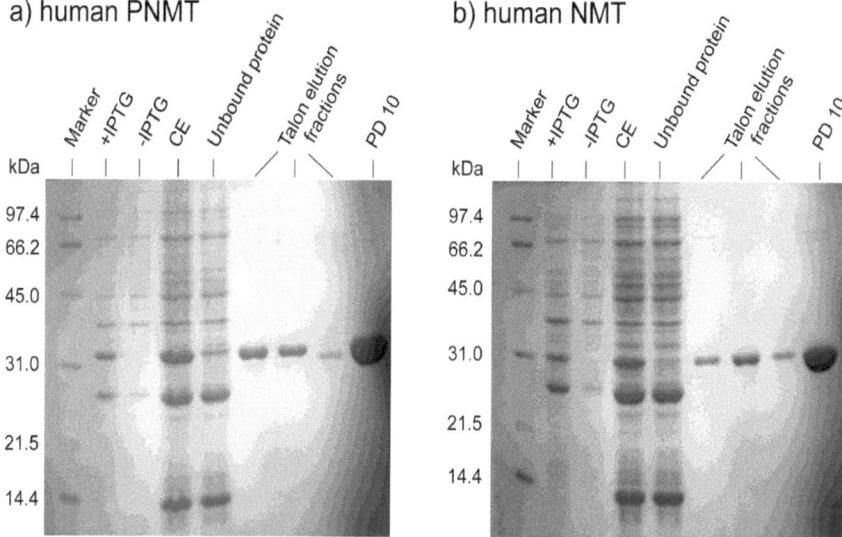

Fig. 28: SDS-PAGE of the purification of human PNMT (clone provided by X. Ren) and human NMT (RZPD) heterologously expressed in *E. coli*. After an induction with IPTG (+-IPTG) a band at 32 kDa can be observed that is missing in the sample taken before the induction (-IPTG). The recombinant protein was found in the crude extract (CE). After incubation with Talon resin the recombinant protein bound and was, therefore, not present in the supernatant of the Talon resin (Unbound protein). The recombinant protein was eluted by high concentrations of imidazol from the resin (Talon elution fractions) and finally desalted (PD10).

a) Steps of purification of the 32 kDa protein human PNMT.
b) Steps of purification of the 32 kDa protein human NMT.

Both His-tagged recombinant proteins were heterologously expressed in *E. coli* according to II.7.3.2. After an induction with IPTG both *N*-methyltransferases were purified from *E.coli* by Talon resin and were desalted *via* PD 10 columns (Fig. 28). In average 10 mg of each 32 kDa protein was harvested from 2 l of *E. coli* culture.

3.2 Activity of Recombinant Phenylethanolamine *N*-Methyltransferase and Amine *N*-Methyltransferase

Enzyme activity of both enzymes was tested with [^{14}C-methyl]-SAM and the substrates that had been described in the literature: phenylethanolamine as substrate for PNMT (AXELROD 1962) and tryptamine as substrate for NMT (THOMPSON *et al.* 1999). Activity of the *N*-methyltransferase was observed when the radio-labeled methyl group of [^{14}C-methyl]-SAM was transferred to the substrate resulting in radio-labeled products that were detected by TLC. After the chiral purity of each of the eight THBIQ enantiomers, that were available in our departmental collection and which are shown in Fig. 29, was verified by chiral separation under the conditions as described in II.10.4.B, they were tested as possible substrates for both proteins. A basic pH optimum of pH 8 had been described for bovine PNMT (CONNET & KIRSHNER 1970) and a pH optimum between pH 8 to 9 had been reported for the two NMT isoforms that were purified from rabbit liver (ANSHER & JAKOBY 1986). These basic conditions were also used in enzyme assays conducted in this thesis to test the activity of both recombinant proteins:

assay PNMT: 1 M potassium phosphate pH 8.0   100 mM
1 mM phenylethanolamine or THBIQ alkaloids (Fig. 29)   10 nmol
0.05 µCi/nmol [^{14}C-methyl]-SAM   0.5 nmol, 0.025 µCi, 50000 cpm
3.5 mg/ml PNMT   200 µg

assay NMT: 1 M Tris HCl pH 8.5   50 mM
20 mM tryptamine or THBIQ alkaloids (Fig. 29)   200 nmol
0.05 µCi/nmol [^{14}C-methyl]-SAM   0.5 nmol, 0.025 µCi, 50000 cpm
3.5 mg/ml NMT   200 µg
ad dd H$_2$O

control assay: boiled protein
incubation: 37°C, 120 min
detection: TLC in solvent system 1 and 6 (II.10.1), analysis with Phosphorimager (II.12)

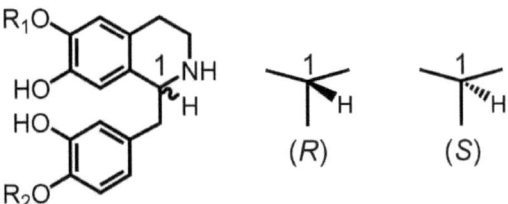

THBIQ	R$_1$	R$_2$
Norlaudanosoline (NLS)	OH	OH
6-O-Methylnorlaudanosoline (6OMeNLS)	CH$_3$	OH
4'-O-Methylnorlaudanosoline (4OMeNLS)	OH	CH$_3$
Norreticuline (NorRet)	CH$_3$	CH$_3$

Fig. 29: (R)- and (S)-THBIQ alkaloids tested as possible substrates for human PNMT and NMT.

Test assays with the THBIQ substrates norlaudanosoline, 6-O-methylnorlaudanosoline and 4'-O-methylnorlaudanosoline were separated in the acidic solvent system 1 (II.10.1). Incubation mixtures with norreticuline as substrate were separated in the ammonium hydroxide containing solvent system 6 (II.10.1). Radioactive spots on each TLC were detected by phosphorimager and quantitated by "ImageQuant" (Molecular Dynamics). PNMT was found to be active with phenylethanolamine but not with any of the offered THBIQ alkaloids (Tab. 11). Enzyme activity tests with NMT revealed that the enzyme was capable of the N-methylation of not only tryptamine but also, most interestingly, exclusively (R)-configurated THBIQ alkaloids (Tab. 11) This high stereospecificity of NMT was observed for all four THBIQ alkaloids that were offered as substrates (Fig. 29).

Whereas the activity of NMT with (R)-6-O-methylnorlaudanosoline and (R)-norlaudanosoline as substrates was rather low, the conversion rate to radio-labeled product increased when (R)-4'-O-methylnorlaudanosoline and (R)-norreticuline were offered as substrates. Incubations were subsequently repeated with both recombinant proteins missing the polyhistidine-tag to exclude a possible interference of the tag with the enzyme activity. The polyhistidine-tag was removed by thrombin digestion as described in II.7.3.3. Both cleaved enzymes verified the results as observed for the His-tagged proteins confirming the lack of activity of PNMT for THBIQ alkaloids and the absolute (R)-specificity of NMT when both enantiomers of THBIQ alkaloids were offered.

3.3 Determination of K_m, V_{max} and the Catalytic Efficiency of NMT with (R)-configurated THBIQ Alkaloids as Substrates

Potassium phosphate buffer at the physiological pH 7.4 was chosen to further analyze the kinetic properties of human NMT. For that, product formation within 10 min of incubation was confirmed to be in a linear range. Following conditions were used to determine the kinetic parameters of human NMT with (R)-configurated THBIQ alkaloids as substrates:

assay: 1 M Tris HCl pH 8.5   50 mM
(R)-THBIQ (Fig. 29)   0.25-2.5 mM, 25-250nmol
1mM SAM   5 nmol
3.5 mg/ml NMT   200 mg

control assay: boiled protein
incubation: 37°C, 10 min
detection: 4000 QTRAP LC-MS with Eclipse HPLC column (II.10.4.A)

Tab. 11: Enzyme activity of human PNMT and human NMT with THBIQ alkaloids. Incubation for 2 h at 37°C revealed that human NMT was active with exclusively (*R*)-configurated THBIQ alkaloids whereas human PNMT was not.
Incubation mixtures consisted of 50 mM Tris/HCl pH 8.5, 0.025 µCi (0.5 nmol, 50000 cpm) [^{14}C-methyl]-SAM, 200 nmol substrate and 200 µg recombinant protein to test the activity of human NMT. Enzyme assays for human PNMT consisted of 100 mM potassium phosphate pH 8.0, 0.025 µCi (0.5 nmol, 50000 cpm) [^{14}C-methyl]-SAM, 20 nmol substrate and 200 µg recombinant protein.

compound	Amine *N*-methyltransferase (EC 2.1.1.49)		Phenylethanolamine *N*-methyltransferase (EC 2.1.1.28)	
	Specific activity (fkat/mg)	Relative efficiency (%)	Specific activity (fkat/mg)	Relative efficiency (%)
(*R*)-6-*O*-Methylnorlaudanosoline	8	8	0	0
(*R*)-Norlaudanosoline	19	20	0	0
(*R*)-4'-*O*-Methylnorlaudanosoline	50	53	0	0
(*R*)-Norrreticuline	94	100	0	0
(*S*)-6-*O*-Methylnorlaudanosoline	0	0	0	0
(*S*)-Norlaudanosoline	0	0	0	0
(*S*)-4'-*O*-Methylnorlaudanosoline	0	0	0	0
(*S*)-Norrreticuline	0	0	0	0

The reaction mixtures were terminated by addition of 10 µl 20% TCA, centrifuged and the supernatant was subjected to LC-MS/MS analysis with an Eclipse XDB HPLC column (II.10.4.A). Michaelis-Menten kinetics plotted by non-linear regression with "Graphpad Prism" are summarized in Fig. 30. Catalytical efficiencies for each substrate were calculated from the estimated Michaelis-Menten parameters (K_m and k_{cat}, see Fig. 30) and revealed that (R)-norreticuline followed by (R)-4'-O-methylnorlaudanosoline were the best substrates for human NMT. The activity and catalytic efficiencies of NMT for (R)-norlaudanosoline and (R)-6-O-methylnorlaudanosoline were much lower. The apparent K_m for (R)-norreticuline was with 0.9 mM three times lower than the one described for tryptamine (THOMPSON et al. 1999) suggesting that the THBIQ alkaloid is a better substrates than the previously claimed natural substrate tryptamine. In addition to the determination of the kinetic parameters for the THBIQ substrates, K_m and k_{cat} for the co-substrate SAM were determined with the best substrate (R)-norreticuline in 2 mM concentration (200 nmol) by incubating the same reaction mixtures as described above but varying the SAM concentration from 0.5-5 nmol (5-50 µM). Compared to THOMPSON et al. (1999) a similar low apparent K_m of 6.9 µM was estimated by non-linear regression with "Graphpad Prism" for the co-subtrate SAM.

The (R)-specific NMT that had no functional role in human physiology might have a new function in morphine biosynthesis. The enzyme was found to catalyze a stereospecific reaction with THBIQ as substrates. This absolute (R)-stereospecificity was not reported before by BAHNMAIER et al. (1999) when they tested (R)- and (S)-configurated TIQ alkaloids as substrates. The results of the enzymatic studies with THBIQ indicate that an addition of a benzyl ring to the TIQ molecule apparently changed the activity to a highly stereospecific N-methyltransferase. The newly observed characteristic of the enzyme to only accept (R)-configurated precursors of mammalian morphine biosynthesis suggested a fundamental difference to the plant pathway in which the initial steps from the trihydroxylated THBIQ norcoclaurine to reticuline are (S)-configurated. In the biosynthesis of morphine in plants, the N-methylation of (S)-coclaurine yields (S)-N-Methylcoclaurine and then after two additional reactions (S)-reticuline that undergoes a change of configuration to (R)-reticuline (Fig. 1).

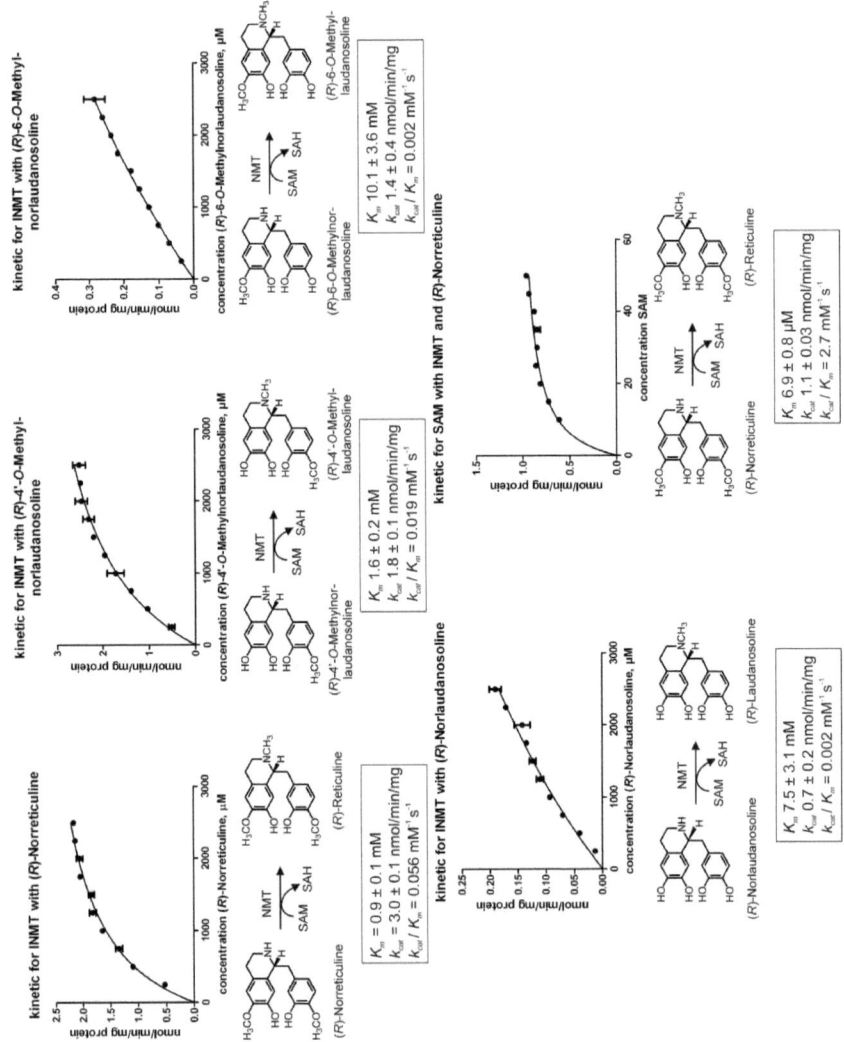

Fig. 30: Non-linear regression and kinetic parameters of NMT for the substrates (R)-norreticuline, (R)-4′-O-methylnorlaudanosoline, (R)-O-methylnorlaudanosoline and (R)-norlaudanosoline and the co-substrate SAM.

assay THBIQ kinetic: 50 mM potassium phosphate pH 7.4, 5 nmol SAM, 25-250 nmol substrate, 200 µg human NMT

assay SAM kinetic: 50 mM potassium phosphate pH 7.4, 0.5-5 nmol SAM, 200 nmol (R)-norreticuline, 200 µg human NMT

incubation: 37°C, 10 min

4 The Oxidative *C-C*-Phenol-Coupling in Mammals

In vivo experiments described in this thesis reported on the urinary excretion of salutaridine after i.p. injection of norlaudanosoline. This finding demonstrated that similar to plants the initial steps of the biosynthesis of morphine in mammals require not only a sequence of methylation reactions but also the conversion of the THBIQ alkaloid into the morphine precursor *via* phenol-coupling. The intramolecular *C-C*-phenol-coupling of the C12-C13 carbon bridge in plant morphine biosynthesis was reported to be catalyzed by salutaridine synthase first purified from *Papaver somniferum* by GERARDY & ZENK (1993b). The highly substrate- and stereo-specific P450 enzyme was recently cloned from an opium poppy cDNA library after its increased expression in morphine producing *Papaver species* was discovered by comparative transcript profiling (GESELL *et al.* 2009). In an attempt to discover the key reaction in morphine biosynthesis in animals uniformly labeled [^{3}H]-racemic reticuline was incubated in the presence of NADPH and rat microsomes and was claimed to be converted to [^{3}H]-labeled salutaridine (WEITZ *et al.* 1987). However, the findings of WEITZ *et al.* (1987) were ambiguous since they used impure, uniformly labeled [^{3}H]-reticuline that gave when incubated with rat liver microsomes a very low conversion rate to salutaridine of only 1%. The experiments were later repeated using [N-^{14}CH$_3$]-(*R*)-reticuline and microsomal bound cytochrome P450 from pig, rat, cow and sheep and formation of [N-^{14}CH$_3$]-salutaridine was observed with the right stereochemistry at carbon 9 (AMANN & ZENK 1991). The enzyme was subsequently purified from porcine liver to homogeneity and the reaction product, characterized by mass spectrometry and physical parameters, was shown to be salutaridine the product of a P450 enzyme named salutaridine synthase (AMMAN *et al.* 1995). The microsome derived P450 enzyme was subjected to Edman degradation and upon comparison of the peptide sequence to those present in the public database homology to human CYP 2D6 was observed. Human CYP 2D6 is a well studied P450 enzyme and became famous for a phenomenon known as the debrisoquine/sparteine metabolism of drug oxidation (reviewed in EICHELBAUM & GROSS 1990). This polymorphism was named after aberrant metabolism patterns, which were discovered for the first prototype drugs debrisoquine (MAHGOUB *et al.* 1977) and sparteine (EICHELBAUM *et al.* 1979), being caused by the absence or deficiency of functional allelic variants of CYP 2D6. At least 112 allelic variants of CYP 2D6 have been described since the discovery of the genetic polymorphism resulting in a total of four individual phenotypes (SANGAR *et al.* 2009). Depending on the activity of CYP 2D6 it is distinguished between poor, intermediate, extensive and ultrarapid metabolizers. Nowadays,

CYP 2D6 is known to be involved in the oxidation of about 30% of drugs used by humans including the morphine precursor codeine (GUENGERICH et al. 2002).

4.1 The Phenol-Coupling Reaction in Mammals Catalyzed by Human CYP 2D6 and CYP 3A4

To answer the question whether CYP 2D6 is capable of catalyzing the *C-C*-phenol-coupling reaction in humans the enzyme was obtained from a commercial source (BD Biosciences). CYP 2D6 was received as a microsomal preparation from baculovirus infected cells expressing CYP 2D6 and coexpressing human cytochrome P450 reductase (CPR), a flavin adenine dinucleotide and flavin mononucleotide containing flavoprotein that functions as an electron donor and is essential for the catalytic activity of P450 enzymes. After the sequence of the P450 enzyme was verified by MS, done by Dr. L. Hicks and Dr. S. Alvarez at the Donald Danforth Plant Science Center, St. Louis, USA, the following reaction mixture was prepared to test the catalytical activity:

assay: 100 mM potassium phosphate pH 6.5   50 µl   (36 mM)
1.37 mM (R)-$[N$-$^{14}CH_3]$-reticuline   5 µl   (6.75 nmol, 0.2 µCi, 250000 cpm)
5mM NADPH   50 µl   (0.25 µmol)
5 mM EDTA   10 µl   (0.05 µmol)
0.056 mg/ml CYP 2D6   20 µl   (*ca.* 1 µg)
                                                ─────
                                                140 µl

control assay: boiled protein
incubation: 37°C, 5 h
detection: 2D-TLC in solvent system 3 and 7 (II.10.1), analysis with Phosphorimager (II.12)

An aliquot of the incubation mixture was subjected to 2D-TLC and four metabolites were detected after incubation of (R)-$[N$-$^{14}CH_3]$-reticuline with CYP 2D6 (Fig. 31a). Since 2D-TLC could not clearly separate metabolite C from (R)-$[N$-$^{14}CH_3]$-reticuline the entire radioactive spot was scratched off and eluted from TLC for re-chromatography in solvent system 8 (II.10.1) in which both compounds were separated (Fig. 31b). Comparison with standards, that were available from our departmental collection and that were expected to be formed according to the phenol coupling reaction, identified metabolite A as corytuberine, metabolite B as pallidine, metabolite C as isoboldine and metabolite D as salutaridine. The possible formation of this set of four products from reticuline by phenol oxidation was already theoretically suggested by BARTON et al. (1967) and KAMETANI & FUKUMOTO (1971) during their investigations of the chemical synthesis of morphinanedienones. Depicted in Fig. 32 are the chemical structures of the four formed products and their biosynthesis from reticuline through C-C-phenol-coupling.

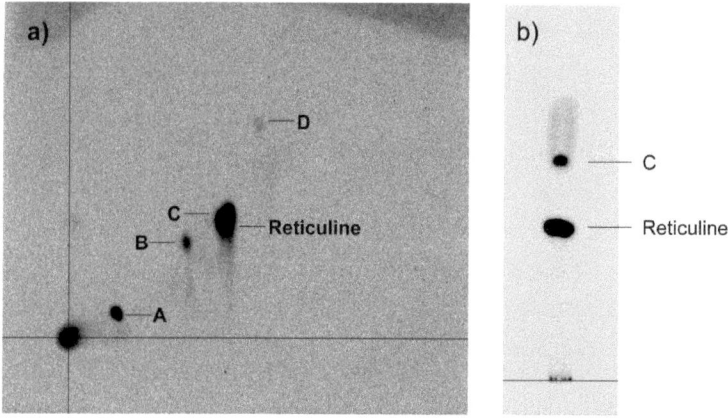

Fig. 31: Four products are formed from [N-$^{14}CH_3$]-(R)-reticuline catalyzed by human CYP 2D6.
assay: 36 mM potassium phosphate buffer pH 6.5, 1 µg CYP 2D6, 1.8 mM NADPH, 0.36 mM EDTA, and 0.2 µCi [N-$^{14}CH_3$]-(R)-reticuline (250000 cpm, 6.75 nmol)
incubation: 37°C, 5 h

a) 2D-TLC of the reaction mixture reveals four unknown metabolites A,B,C and D produced from [N-$^{14}CH_3$]-(R)-reticuline *via* oxidative phenol coupling catalyzed by CYP 2D6.
b) Rechromatography of the spot in Fig. 31a) containing metabolite C and [N-$^{14}CH_3$]-(R)-reticuline.

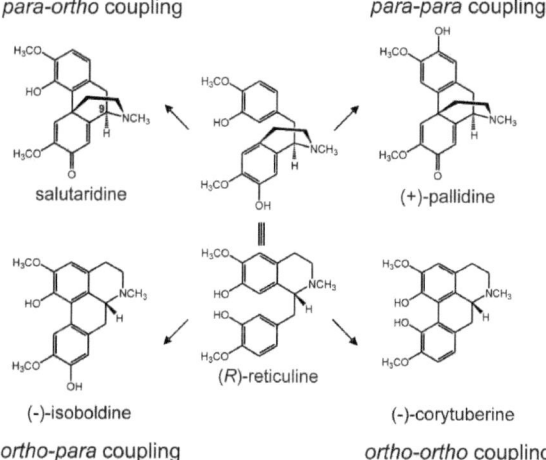

Fig. 32: Oxidative phenol-coupling reaction of (R)-reticuline in mammals. The four phenol-coupled products salutaridine, (+)-pallidine, (-)-isoboldine and (-)-corytuberine were formed from (R)-reticuline *via* an oxidative phenol-coupling reaction catalyzed by human CYP 2D6.

To verify the formation of the four products by LC-MS the same enzyme assay as for the radioactive detection was prepared except that unlabeled (*R*)-reticuline was used. Alkaloids were extracted for 10 min by adding 400 µl sodium carbonate pH 9.5 and 400 µl chloroform to the reaction mixture. The supernatant was evaporated and resuspended in methanol and subjected to LC-MS with a Luna HPLC column (II.10.4.C). Comparison with standards of alkaloids that were expected to be formed according to the phenol coupling reactions and that were identified by TLC verified the products unequivocally by MS and retention time as (-)-corytuberine (*ortho-ortho* coupling), (+)-pallidine (*para-para* coupling), (-)-isoboldine (*ortho-para* coupling) and salutaridine (*para-ortho* coupling) (Fig. 33). The identification of these compounds is consistent with previous findings with rat liver microsomes and racemic reticuline described by KAMETANI *et al.* (1977, 1980). However, the important morphine precursor salutaridine was missed during that investigation.

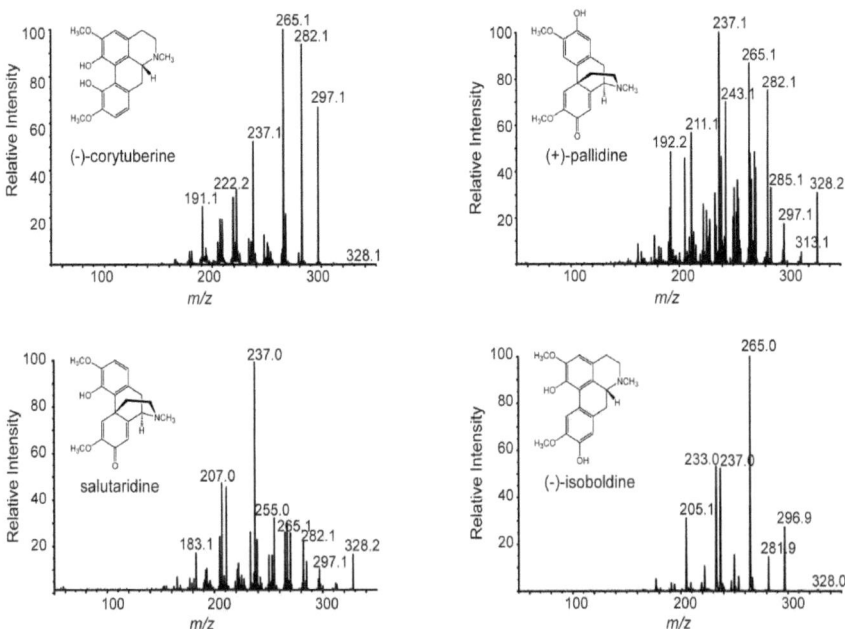

Fig. 33: CID spectra, obtained with a 4000 QTRAP, for the four phenol-coupled products formed from (*R*)-reticuline catalyzed by CYP 2D6..
    assay: 36 mM potassium phosphate buffer pH 6.5, 1 µg CYP 2D6, 1.8 mM NADPH, 0.36 mM EDTA and 50 µM (*R*)-reticuline.
    incubation: 37°C, 2 h

A second human P450 enzyme CYP 3A4 that is involved in the oxidation of 50% of drugs and xenobiotics used nowadays by humans has previously been shown to N-demethylate codeine to N-norcodeine (CARACO et al. 1996, YUE & SÄWE 1997). Because of the similarity of CYP 3A4 in accepting codeine as substrate the enzyme was also obtained from a commercial source (BD Biosciences) and tested for its ability to catalyze the phenol coupling of (R)-reticuline. The same enzyme assay as described for CYP 2D6 was prepared for LC-MS/MS detection with unlabeled (R)-reticuline as substrate. As depicted in Fig. 34 CYP 3A4 showed the product pattern known from CYP 2D6 that is the formation of

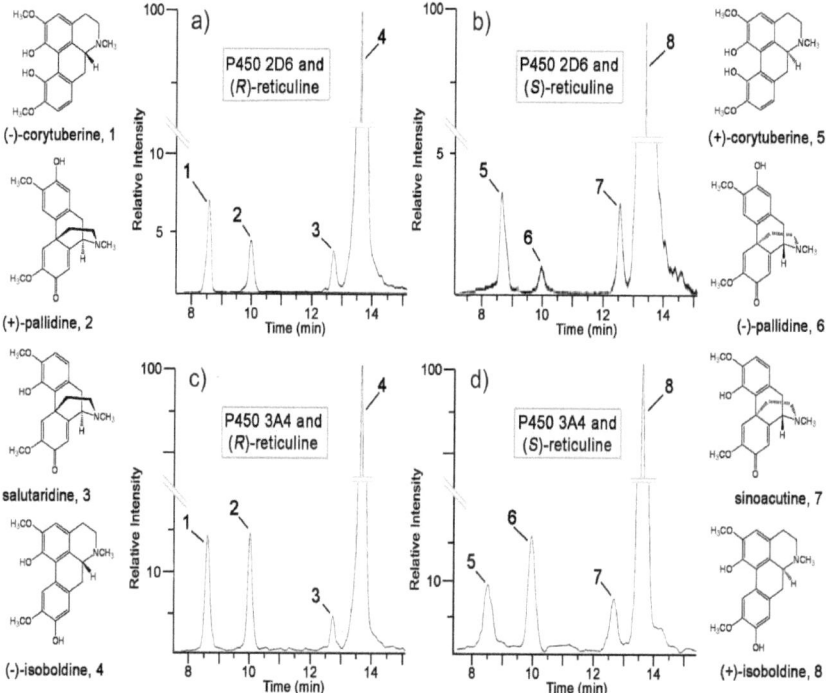

Fig. 34: LC-MS analysis of products formed by human CYP 2D6 and CYP 3A4. The MRM transition from m/z 328 to m/z 237 showed the product formation after 2 h of incubation.
    assay: 36 mM potassium phosphate buffer pH 6.5, 1 µg P450 enzyme, 1.8 mM NADPH, 0.36 mM EDTA and 50 µM (R)-reticuline.
    incubation: 37°C, 2 h

a) CYP 2D6 and (R)-reticuline
b) CYP 2D6 and (S)-reticuline
c) CYP 3A4 and (R)-reticuline
d) CYP 3A4 and (S)-reticuline

salutaridine, (+)-pallidine, (-)-isoboldine and (-)-corytuberine from (R)-reticuline as substrate as determined by MS. Subsequently, the stereospecificity of both enzymes CYP 2D6 and CYP 3A4 was tested by preparing the same assay as described before differing only in that (S)-reticuline was used as substrate. As shown in Fig. 34 four (S)-configured products were detected by LC-MS/MS (II.10.4.C). They were identified as sinoacutine, (-)-pallidine, (+)-isoboldine and (+)-corytuberine. Similar yields for the (R)- and (S)-series of the phenol coupled products were detected regardless of the P450 catalyst chosen, suggesting that the substrates are oxidized without stereoselectivity. In order to elucidate if the phenol-coupled alkaloid salutaridine produced by these two mammalian enzymes can be converted into thebaine during the plant morphine biosynthesis, feeding experiments on poppy plants that enrich the alkaloid thebaine were conducted. Incubation mixtures with the substrates [N-$^{14}CH_3$]-labeled (R)- and (S)-reticuline were prepared as described above and separated by TLC in solvent system 3. Corresponding bands for [N-$^{14}CH_3$]-salutaridine and [N-$^{14}CH_3$]-sinoacutine were eluted with methanol and reconstituted in water. The *para-ortho* coupled products were fed to five-day-old poppy seedlings as described in II.6.1 and unequivocally verified salutaridine by incorporation into thebaine (Fig. 35).

4.2 The Phenol-Coupling Reaction in Mammals Catalyzed by Rat CYP 2D2

The capability of liver microsomes, obtained from Dark-Agouti rats lacking CYP 2D1, to *O*-demethylate codeine was compared with liver microsomes obtained from Sprague-Dawley rats that expressed active CYP 2D1 (MIKUS *et al.* 1991a). Since the researchers found a 10-fold decrease of morphine formation from codeine in Dark-Agouti rat liver microsomes they concluded CYP 2D1 is the biocatalyst of that 3-*O*-demethylation reaction. This hypothesis was supported later on by immunoinhibition and chemical-inhibitor studies revealing that anti CYP 2D1 antibodies inhibited codeine-*O*-demethylation of rat liver microsomes (XU *et al.* 1997). Although no direct enzyme assay demonstrated that rat CYP 2D1 could be the homolog of human CYP 2D6 and catalyze the 3-*O*-demethylation of codeine, the hypothesis prompted us to investigate if rat CYP 2D1 might be capable to catalyze the phenol-coupling reaction of (R)-reticuline. After rat CYP2D1 was obtained from a commercial source (BD Biosciences), the sequence of the P450 enzyme was verified by MS. The enzyme was subsequently tested for its phenol-coupling activity with (R)-[N-$^{14}CH_3$]-reticuline as substrate for radioactive detection as well as with unlabeled (R)-reticuline for LC-MS detection in the same reaction mixtures as described for human CYP 2D6 and CYP 3A4.

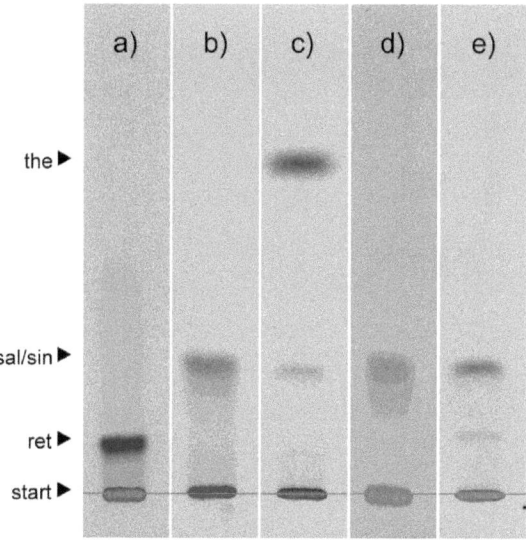

Fig. 35: TLC-Radiogram of feeding of mammalian phenol-coupled alkaloids to *P. somniferum* seedlings; the, thebaine; sal, salutaridine; sin, sinoacutine; ret, reticuline.

a) ^{14}C-labeled reticuline standard
b) ^{14}C-labeled salutaridine formed by enzymatic phenol-coupling of (*R*)-[*N*-^{14}CH$_3$]-reticuline
c) ^{14}C-labeled thebaine in extract of *Papaver* seedlings fed with ^{14}C-labeled salutaridine shown in B
d) ^{14}C-labeled sinoacutine formed by enzymatic phenol-coupling of (*S*)-[*N*-^{14}CH$_3$]-reticuline
e) extract of *Papaver* seedlings fed with ^{14}C-labeled sinoacutine shown in D. Sinoacutine was not incorporated into thebaine, an intermediate of plant morphine biosynthesis.

Surprisingly, the rat P450 enzyme was found to be inactive with (*R*)- or (*S*)-reticuline as substrate (Fig. 36a,b) as well as in additional enzyme tests with codeine as substrate. Based on these findings it was assumed that CYP 2D1 was incorrectly suggested to be the catalyst for the conversion of codeine to morphine. Subsequently, the rat genome was searched for isoformes of CYP 2D1. Five P450 functional forms were found: CYP 2D2, CYP 2D3, CYP 2D4, CYP 2D5 and CYP 2D18 with an identity to CYP 2D1 of 73%, 79%, 72%, 95% and 72%, respectively. Among all these isoforms, CYP 2D2 was described in literature to be poorly expressed in the Dark-Agouti rat strain next to CYP 2D1, too (YAMAMOTO et al. 1998). In addition, rat CYP 2D1 showed similar catalytic activities like human CYP 2D6 (HIROI et al. 2002). It was concluded in this thesis that CYP 2D2 was qualified to the same extent as CYP 2D1 to cause the deficiency of Dark-Agouti rat liver microsomes to metabolize codeine.

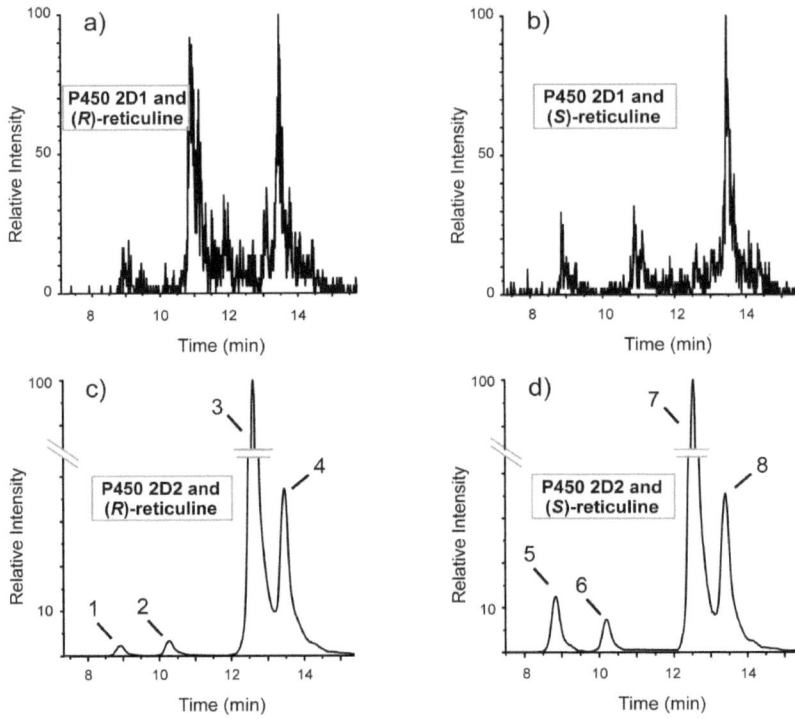

Fig. 36: LC-MS analysis of the incubation of rat CYP 2D1 and rat CYP 2D2 with (R)-reticuline. Shown is the MRM transition from m/z 328 to m/z 237 that revealed product formation for rat CYP 2D2 but not rat CYP 2D1. 1, (-)-corytuberine; 2, (+)-pallidine; 3, salutaridine; 4, (-)-isoboldine; 5, (+)-corytuberine; 6, (-)-pallidine; 7, sinoacutine; 8, (+)-isoboldine
assay: 36 mM potassium phosphate buffer pH 6.5, 50 µM (R)-reticuline, 1.8 mM NADPH, 0.36 mM EDTA and 1 µg CYP 2D1 or 1 µg CYP 2D1.
incubation: 37°C, 2 h

a) CYP 2D1 and (R)-reticuline
b) CYP 2D1 and (S)-reticuline
c) CYP 2D2 and (R)-reticuline
b) CYP 2D2 and (S)-reticuline

To test this hypothesis rat CYP 2D2 was obtained from a commercial source (BD Biosciences). After the sequence of the P450 enzyme was verified by MS, an incubation with codeine as substrate was conducted and the P450 enzyme was found to be active. The 3-O-demethylation reaction of codeine to morphine catalyzed by rat CYP 2D2 is discussed in detail in III.C.5. After this activity was confirmed, rat CYP 2D2 was immediately tested for its capability to catalyze the phenol-coupling reaction of (R)-reticuline. The same incubation

Tab. 12: Relative distribution of phenol-coupled products formed from incubation of (R)- and (S)-reticuline with CYP 2D6, CYP 3A4 and CYP 2D2.

enzyme	(-)-corytuberine / (+)-corytuberine (%)	(+)-pallidine / (-)-pallidine (%)	salutaridine / sinoacutine (%)	(+)-isoboldine / (-)-isoboldine (%)
CYP 2D6	4 / 2	23 / 17	7 / 5	66 / 76
CYP 3A4	8 / 12	22 / 17	34 / 35	36 / 36
CYP 2D2	1 / 4	20 / 34	74 / 56	5 / 4

mixture as described previously for human CYP 2D6 and human CYP 3A4 was prepared. The analysis by LC-MS (II.10.4.C) demonstrated that CYP 2D2 was catalytically active and the same set of four phenol-coupled products (-)-corytuberine, (+)-pallidine, salutaridine and (-)-isoboldine was formed (Fig. 36c). The phenol-coupling reaction catalyzed by CYP 2D2 was found to be not stereospecific like it was shown for human CYP 2D6 and human CYP 3A4 since all four (S)-configured products (+)-corytuberine, (-)-pallidine, sinoacutine and (+)-isoboldine were identified, too (Fig. 36d). Rat CYP 2D2 seemed to be the most effective catalyst for the production of salutaridine from (R)-reticuline compared to the two human P450 enzymes.The relative distribution of all four phenol-coupled products formed from (R)- and (S)-reticuline by the action of human CYP 2D6, human CYP 3A4 and rat CYP 2D2 is summarized in Tab. 12 and was shown to be the same for all reactions independent of pH, enzyme concentration and incubation time.

4.3 Kinetic Analysis and Characteristics of Mammalian P450 Enzymes Catalyzing the Phenol Coupling Reaction

The P450 enzymes human CYP 2D6, human CYP 3A4 and rat CYP 2D2 that were obtained from BD Biosciences expressed as functional protein from baculovirus infected insect cells and that were all shown to be capable of catalyzing the phenol-coupling reaction, were characterized in regard to their pH optima. Enzyme reactions with all three P450 enzymes were carried out according to HANNAH et al. (2001). Additionally, CYP 3A4 was analyzed in presence of coexpressed cytochrome b_5 since catalytical activities of CYP 3A4, but not of CYP 2D6, were shown to be enhanced by addition of cytochrome b_5 that is possibly involved in facilitating electron transfer (YAMAZAKI et al. 2002). To obtain pH optimum curves the following incubation mixtures were prepared:

assay:	1 M buffer	25 µl	(100 mM)
	100 µM (R)-reticuline	25 µl	(2.5 nmol)
	2 mg/ml DLPC	7.5 µl	(15 µg)
	20 mM NADP$^+$	12.5 µl	(250 nmol)
	100 mM glucose-6-phosphate	25 µl	(2.5 µmol)
	50 units/ml glucose-6-phosphate dehydrogenase	5 µl	(0.25 units)
	100 units/µl catalase	5 µl	(500 units)
	1 mg/ml superoxide dismutase	10 µl	(10 µg)
	0.056 mg/ml CYP 2D6	18 µl	(*ca.* 1 µg)
or	0.057 mg/ml CYP 3A4	9 µl	(*ca.* 0.5 µg)
or	0.057 mg/ml CYP 3A4 + cytochrome b_5	9 µl	(*ca.* 0.5 µg)
or	0.056 mg/ml CYP 2D2	5 µl	(*ca.* 0.3 µg)
		250 µl	

buffer: citrate/NaOH pH 4-6, phosphate pH 6-8, Tris/HCl pH 8-10
incubation: 37°C, 10 min
detection: 4000 QTRAP LC-MS with Eclipse HPLC column (II.10.4.C)

After incubation of the samples, 400 µl 1 M sodium carbonate buffer pH 9.5 were added and the reaction mixtures were extracted twice with 400 µl chloroform for 10 min followed by a centrifugation to separate the organic from the aqueous phase. The combined organic phases were dried, reconstituted with methanol and subjected to LC-MS/MS with a Luna HPLC column (II.10.4.C). Surprisingly, human CYP 2D6 showed an acidic pH optimum at pH 5.5 (Fig. 37a), while CYP 3A4 with or without cytochrome b_5 and CYP 2D2 showed an optimum around the physiological pH 7.4 (Fig. 37b,c,d). The finding of an acidic pH optimum for human CYP 2D6 was inconsistent with earlier reports on CYP 2D6 describing an optimal pH of 7-7.5 for the hydroxylation of bufuralol, another CYP 2D6 substrate (OSCARSON *et al.* 1997). Subsequently, we received for further analyses *E. coli* expressed CYP 2D6 from our collaborator, Prof. F.P. Guengerich (Vanderbilt University School of Medicine, Nashville, TN, USA). The same incubation mixtures as described for the BD Biosciences enzyme was conducted except that 38 pmol human CYP 2D6 and 255 pmol rat CPR, that had to be added exogenously, were used. As can be seen in Fig. 37e human CYP 2D6 obtained from our collaborating lab showed a different pH optimum of pH 6.5. The acidic pH optimum of pH 5.5 for the BD Biosciences enzyme was inexplainable particularly after the sequence of the purchased protein, human CYP 2D6, was verified by the published amino acid sequence.

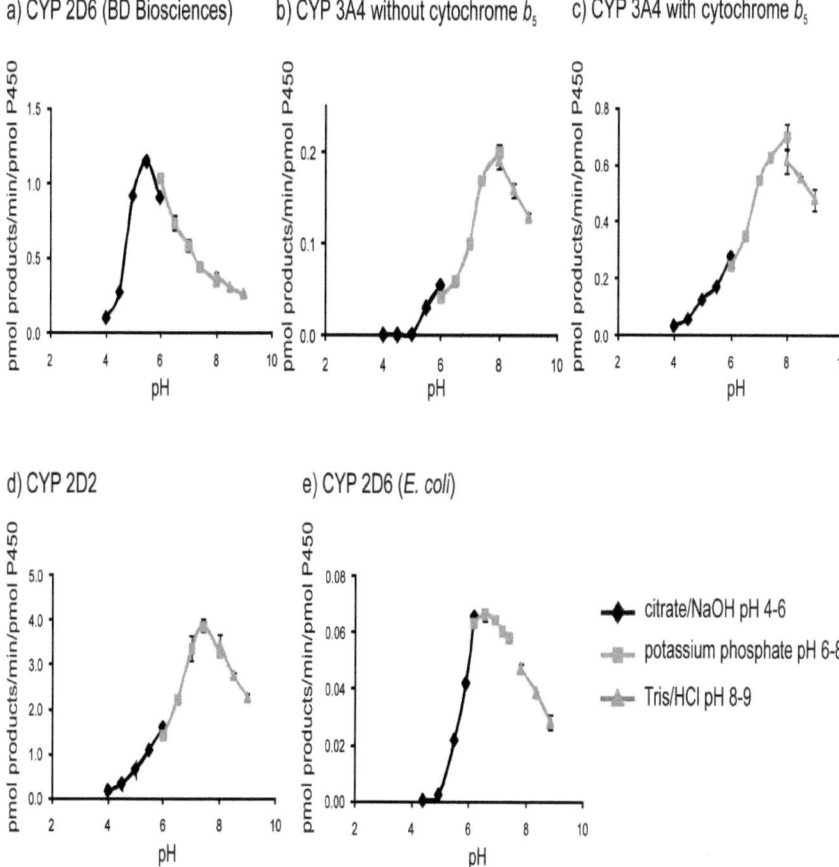

Fig. 37: pH-optimum curves for mammalian P450 enzymes catalyzing the phenol-coupling reaction of (R)-reticuline to a set of four products. Illustrated is the total amount of all four products. The pH corresponds to the pH measured in the reaction mixture.

    assay: 100 mM buffer, 2.5 nmol (R)-reticuline, 15 µg DLPC, 250 nmol $NADP^+$, 2.5 µmol glucose-6-phosphate, 0.25 units glucose-6-phosphate dehydrogenase, 500 units catalase, 10 µg superoxide dismutase, 0.3-1 µg P450 enzyme.

    incubation: 37°C, 10 min

a) pH-Optimum of CYP 2D6 from BD Biosciences.
b) pH-Optimum of CYP 3A4 from BD Biosciences without co-expressed cytochrome b_5.
c) pH-Optimum of CYP 3A4 from BD Biosciences with co-expressed cytochrome b_5.
d) pH-Optimum of CYP 2D2 from BD Biosciences.
e) pH-Optimum of CYP 2D6 from Prof. F.P. Guengerich expressed in *E. coli*.

A direct comparison of activity between human CYP 2D6 obtained from BD Biosciences with the protein from our collaborating lab (Fig. 37a,e) revealed that the *E. coli* expressed protein produced 10 times less phenol-coupled product at the physiological pH 7.4 than the protein expressed in insect cells. The expression system *per se* might be important for the activity of eukaryotic P450 enzymes. Insect cells provide a membrane system that could be a necessary requirement for the expression of P450 enzymes in the right conformation. Also, co-expression of human CPR in the insect cell expression system could have been a reason for the increase in P450 activity. Comparison of the activities of two CYP 3A4 expression systems (Fig. 37b,c) revealed that addition of cytochrome b_5 increased product formation by almost four times confirming the observations by YAMAZAKI *et al.* (2002). The role of cytochrome b_5 and particularly why it stimulates some CYP 3A4 reactions and some not, is however, still unclear.

To determine the kinetic characteristics of all P450 enzymes catalyzing the mammalian phenol-coupling reaction, incubations were carried out under the conditions as described for the pH-optimum in 100 mM potassium phosphate pH 7.4 by varying the concentration of reticuline between 0.25 µM to 5 mM depending on the biocatalyst. Michaelis-Menten kinetics were plotted by non-linear regression with "GraphPad Prism" and corresponding curves are shown in Fig. 38.

From their Michaelis-Menten parameters distinct kinetics were derived listed in Tab. 13 for the phenol-coupling reaction catalyzed by human CYP 2D6, rat CYP 2D2 and human CYP 3A4 in absence or presence of cytochrome b_5. CYP 2D6 and CYP 2D2 had similar low apparent K_m of 2-3 µM and 0.5-1.1 µM, respectively, for the substrate (*R*)-reticuline as compared to CYP 3A4 (K_m = 500-1000 µM) whereas a higher maximum rate (k_{cat}) for the conversion of (*R*)-reticuline was observed for CYP 3A4. Including cytochrome b_5 in the CYP 3A4 system increased the rate of product formation (k_{cat}) by about 6 to 7 times whereas the K_m was only slightly changed. Catalytic efficiencies for both human enzymes, CYP 2D6 and CYP 3A4 utilizing (*R*)-reticuline as a substrate, showed that CYP 2D6 seemed to be more efficient; however, if cytochrome b_5 was added to the CYP 3A4 system, catalytic efficiencies of both enzymes became comparable. Overall, rat CYP 2D2 showed the highest catalytic efficiency compared to the two human P450 enzymes.

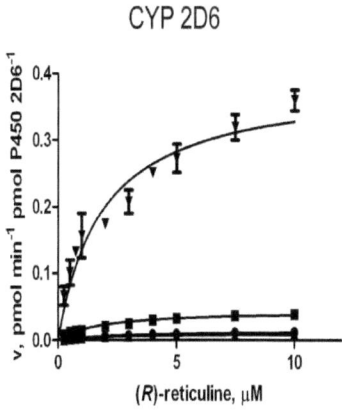
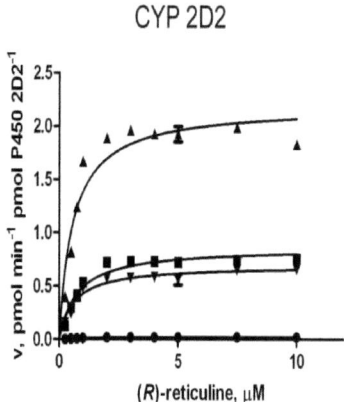
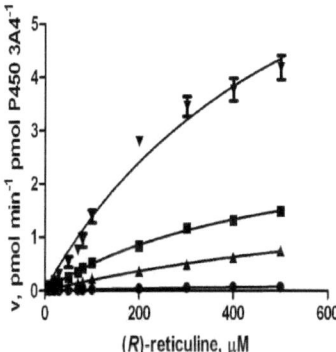
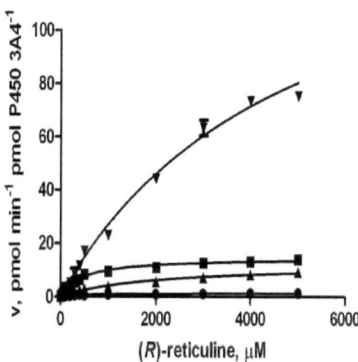

Fig. 38: Michaelis-Menten kinetics derived by non-linear regression for the product formation of salutaridine (▲), (+)-pallidine (■), (-)-isoboldine (▼) and (-)-corytuberine (●) from (R)-reticuline by human CYP 2D6, rat CYP 2D2 and human CYP 3A4 in absence or presence of cytochrome b_5.

assay: 100 mM potassium phosphate pH 7.4, 0.25 μM – 5 mM (R)-reticuline, 15 μg DLPC, 250 nmol $NADP^+$, 2.5 μmol glucose-6-phosphate, 0.25 units glucose-6-phosphate dehydrogenase, 500 units catalase, 10 μg superoxide dismutase, 0.3-1 μg P450 enzyme.

incubation: 37°C, 10 min

Tab 13: Catalytic parameters for the phenol-coupling reaction catalyzed by human CYP 2D6, rat CYP 2D2 and human CYP 3A4 in the presence or absence of cytochrome b_5 (incubation conditions as in Fig. 38).

Enzyme	Product	K_m, µM	k_{cat}, pmol/min/pmol P450	k_{cat}/K_m, mM^{-1} s^{-1}
CYP 2D6	Corytuberine	2.7±0.37	0.015±0.001	0.09
	Pallidine	1.8±0.26	0.045±0.002	0.42
	Salutaridine	2.5±0.21	0.011±0.001	0.08
	Isoboldine	1.9±0.36	0.390±0.026	3.42
CYP 2D2	Corytuberine	1.0±0.18	0.025±0.001	0.42
	Pallidine	0.8±0.10	0.860±0.029	17.9
	Salutaridine	0.7±0.12	2.262±0.107	53.9
	Isoboldine	0.6±0.10	0.693±0.025	19.3
CYP 3A4 with cytochrome b_5	Corytuberine	384±40	1.3±0.1	0.06
	Pallidine	474±30	14.7±0.3	0.52
	Salutaridine	1961±181	12.3±0.5	0.12
	Isoboldine	4860±860	158±17	0.54
CYP 3A4	Corytuberine	534±66	0.16±0.14	0.005
	Pallidine	490±38	2.97±0.14	0.101
	Salutaridine	993±220	2.20±0.36	0.037
	Isoboldine	577±130	9.37±1.34	0.271

5 3-*O*-Demethylation of Thebaine and Codeine in Mammals Catalyzed by Human CYP 2D6 and Rat CYP 2D2

Human CYP 2D6 has been described to catalyze the 3-*O*-demethylation of codeine to morphine (DAYER et al. 1988, CHEN et al. 1988) and predicted to convert thebaine to oripavine (MIKUS et al. 1991a). To test whether rat CYP 2D2 and human CYP 3A4, that were shown to catalyze the phenol-coupling reaction in morphine biosynthesis in mammals, would also be capable of the 3-*O*-demethylation of codeine and thebaine following reaction mixtures were prepared:

assay:
1 M potassium phosphate pH 7.4	25 µl	(100 mM)
100 µM thebaine or codeine	25 µl	(2.5 nmol)
2 mg/ml DLPC	7.5 µl	(15 µg)
20 mM NADP$^+$	12.5 µl	(250 nmol)
100 mM glucose-6-phosphate	25 µl	(2.5 µmol)
50 units/ml glucose-6-phosphate dehydrogenase	5 µl	(0.25 units)
100 units/µl catalase	5 µl	(500 units)
1 mg/ml superoxide dismutase	10 µl	(10 µg)
or 0.057 mg/ml CYP 3A4	9 µl	(ca. 0.5 µg)
or 0.057 mg/ml CYP 3A4 + cytochrome b_5	9 µl	(ca. 0.5 µg)
or 0.056 mg/ml CYP 2D2	5 µl	(ca. 0.3 µg)
	250 µl	

control assay: without P450 enzyme
incubation: 37°C, 2 h
detection: 4000 QTRAP LC-MS with Eclipse HPLC column (II.10.4.C)

After the incubation, 400 µl 1M sodium carbonate buffer pH 9.5 were added to the samples and the reaction mixtures were extracted twice with 400 µl chloroform for 10 min. A centrifugation separated the organic from the aqueous phase. The combined organic phases were dried, reconstituted with methanol and subjected to LC-MS/MS with an Eclipse XDB-C18 HPLC column (II.10.4.C). Only CYP 2D2 was capable of converting thebaine into oripavine and codeine into morphine (Fig. 39). Subsequently, reaction mixtures as described above containing 0.3 µg rat CYP 2D2 were incubated for 10 min with 2.5-50 µM thebaine or codeine to obtain Michaelis-Menten kinetic parameters estimated by non-linear regression with "GraphPad Prism" (Fig. 40). Reaction mixtures as described above with 2.5-50 µM thebaine as substrate and 0.3 µg CYP 2D6 were incubated to compare kinetic parameters of rat CYP 2D2 with the human homolog. For the conversion of codeine to morphine by human CYP 2D6 values for K_m and k_{cat} were taken from literature (OSCARSON et al. 1997, YU et al. 2002).

As summarized in Tab. 14 it was clearly found that the rat homolog CYP 2D1 was catalytically more efficient than the human P450 enzyme CYP 2D6 in catalyzing the 3-O-demethylation of thebaine and codeine. The apparent K_m values for the two reactions of CYP 2D2 were comparable (K_m = 27 µM and 28 µM) with the K_m value of the 3-O-demethylation of thebaine catalyzed by CYP 2D6 (K_m = 37 µM). Only the K_m for the 3-O-demethylation of codeine catalyzed by CYP 2D6 was slightly higher (K_m = 190/250 µM). Compared to kinetic parameters for the phenol-coupling reaction shown in Tab. 13 the 3-O-demethylation proceeded generally faster (higher k_{cat}) but with a higher K_m.

a) Oripavine standard

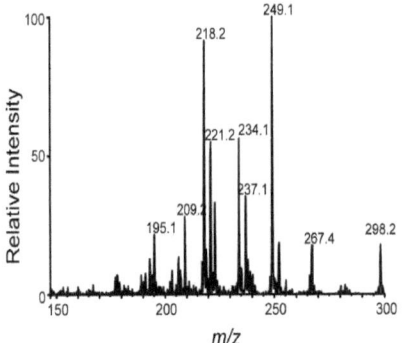

b) Oripavine as enzymatic product

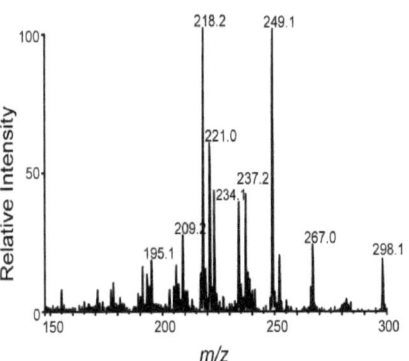

c) Morphine standard

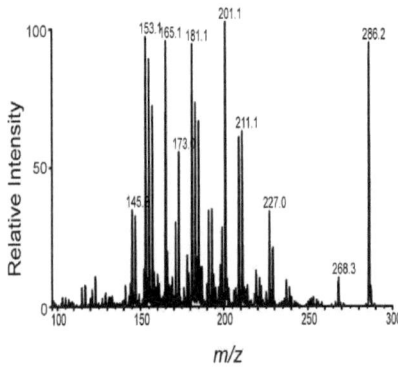

d) Morphine as enzymatic product

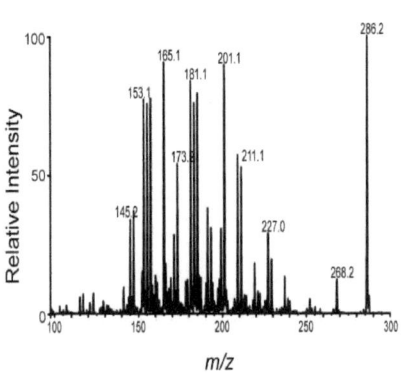

Fig. 39: CID spectra, obtained with a 4000 QTRAP, for oripavine formed from thebaine and for morphine formed from codeine catalyzed by rat CYP 2D2.

assay: 100 mM potassium phosphate pH 7.4 , 2.5 nmol substrate thebaine or codeine, 15 µg DLPC, 250 nmol NADP$^+$, 2.5 µmol glucose-6-phosphate, 0.25 units glucose-6-phosphate dehydrogenase, 500 units catalase, 10 µg superoxide dismutase, 0.3 µg CYP 2D2.

incubation: 37°C, 2 h

a) CID spectra of oripavine standard.
b) CID spectra of oripavine formed from thebaine catalyzed by CYP 2D2.
c) CID spectra of morphine standard.
d) CID spectra of morphine formed from codeine catalyzed by CYP 2D2.

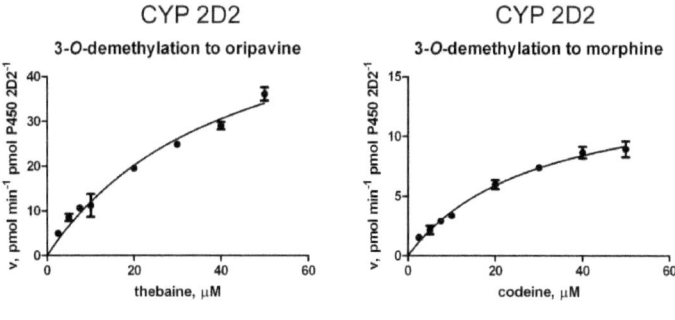

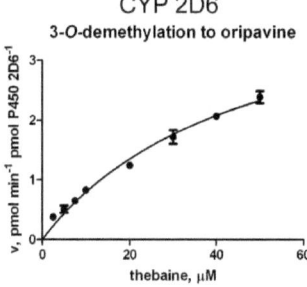

Fig. 40: Michaelis-Menten kinetics derived by non-linear regression for the product formation of oripavine from thebaine and morphine from codeine by rat CYP 2D2 and human CYP 2D6.
assay: 100 mM potassium phosphate pH 7.4 , 0.25 – 50 µM thebaine, 15 µg DLPC, 250 nmol NADP$^+$, 2.5 µmol glucose-6-phosphate, 0.25 units glucose-6-phosphate dehydrogenase, 500 units catalase, 10 µg superoxide dismutase, 0.3 µg P450 enzyme.
incubation: 37°C, 10 min

Tab. 14: Catalytic parameters for the 3-*O*-demethylation of thebaine and codeine catalyzed by human CYP 2D6 and rat CYP 2D2.

Enzyme	Substrate	Product	K_m, µM	k_{cat}, pmol/min/pmol P450	k_{cat}/K_m, mM^{-1} s^{-1}
CYP 2D6	Thebaine	Oripavine	48±9.6	4.6±0.5	1.60
	Codeine	Morphine	250[a]	14[a]	0.93
			190[b]	6.4[b]	0.56
CYP 2D2	Thebaine	Oripavine	42±9.0	63±7.6	25.0
	Codeine	Morphine	29±4.6	14±1.1	8.0

a) OSCARSON *et al.* (1997), b) YU *et al.* (2002).

6   6-*O*-Demethylation of Thebaine and Oripavine

In an incubation mixture with rat liver microsomes and an NADPH-generating system, the cofactor NADH was found to be critical for the 6-*O*-demethylation reaction of thebaine to codeine and oripavine to morphine, while the 3-*O*-demethylation of thebaine to oripavine and codeine to morphine did not require this cofactor (KODAIRA & SPECTOR 1988, FISINGER 1998). The NADH-dependency indicated the presence of an enzymatic system reducing of codeinone to codeine and morphinone to morphine possibly catalyzed by morphinone reductase (TODAKA et al. 2000). However, the vinyl ether splitting enzyme is unknown both in plants and animals so far. To test whether orphan P450 enzymes could be involved in the 6-*O*-demethylation of thebaine and oripavine in animals it was suggested to incubate liver microsomes with P450 inhibitors (personal communication M.H. Zenk/F.P. Guengerich). For that, a sensitive analytical method is required that is capable to identify minimal quantities of the products. The development of such a sensitive method was attempted within this thesis. According to II.9 rat liver microsomes were incubated with [*N*-CD$_3$]-thebaine as substrate in the presence of an NADPH-generating system and the cofactor NADH as described by KODAIRA & SPECTOR (1988). The incubation mixtures were terminated by adding 100 µl 10 % TCA. After centrifugation the supernatant was subjected to LC-MS/MS analysis with an Eclipse XDB-C18 HPLC column (II.10.4.A). Rat liver microsomes clearly converted the substrate into [*N*-CD$_3$]-codeine by 6-*O*-demethylation of [*N*-CD$_3$]-thebaine which was then further 3-*O*-demethylated by CYP 2D2/ CYP 2D6 yielding [*N*-CD$_3$]-morphine (Fig. 41a). These two products were not detected in control assays with boiled protein. Since the amounts were close to the LOD a solid phase extraction was conducted to further concentrate the alkaloids in the incubation mixtures (II.9.4). After removal of protein by TCA the pH of the reaction mixture was adjusted to 8-8.5. The sample was loaded onto a preconditioned Sep Pak Plus C18 cartridge (Varian) which was rinsed and eluted with 5 ml 0.5% ammonium hydroxide in methanol. The concentrated sample was subjected to LC-MS/MS and as depicted in Fig. 41b about 10 times more [*N*-CD$_3$]-codeine and [*N*-CD$_3$]-morphine were detected than without previous SPE. Subsequently, this analytical test system was used to verify the results of KODAIRA & SPECTOR (1988) and FISINGER (1998) that addition of NADH to the reaction mixture with thebaine as substrate leads to an increase of the 6-*O*-demethylated product codeine as it is clearly shown in Fig. 41c. Additionally, thebaine is also 3-*O*-demethylated to oripavine by rat CYP 2D2/ human CYP 2D6 that after addition of NADH is further 6-*O*-demethylated by a yet unknown P450 enzyme to morphine (Fig. 41c).

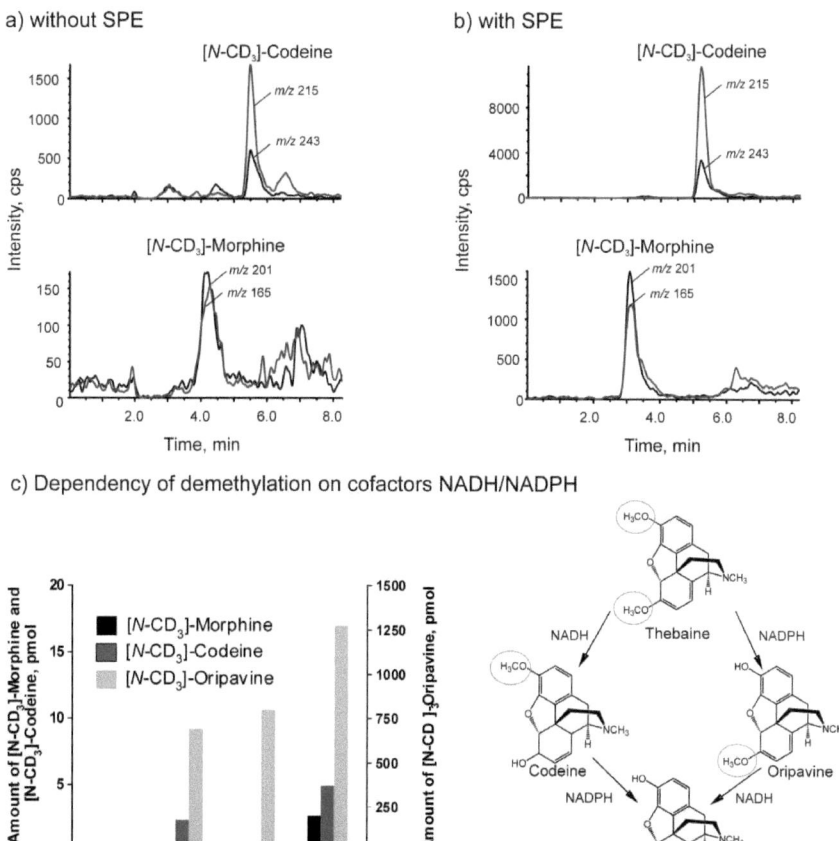

Fig. 41: Experiments on the 6-O-demethylation of thebaine by rat liver microsomes.
assay: 50 mM potassium phosphate pH 7.4 , 10 nmol [N-CD$_3$]-thebaine, 5.5 µmol glucose-6-phosphate, 1 unit glucose-6-phosphate dehydrogenase, 1 µmol NADP$^+$, 5 µmol MgCl$_2$, 1 µmol NADH, 3 mg rat liver microsomes.
incubation: 37°C, 120 min

a) MRM transition for the enzymatic products [N-CD$_3$]-codeine (m/z 303 → 215, m/z 303 → 243) and [N-CD$_3$]-morphine (m/z 289 → 201, m/z 289 → 165).
b) MRM transition for the enzymatic products [N-CD$_3$]-codeine (m/z 303 → 215, m/z 303 → 243) and [N-CD$_3$]-morphine (m/z 289 → 201, m/z 289 → 165) after concentration by SPE.
c) Distinct cofactor dependencies for the 6-O-demethylation and 3-O-demethylation are suggested for the bifurcate pathway in animals.

IV Discussion

Early predictions in the beginning of the 20th century about the existence of opioid receptors mediating pharmacological effects were verified biochemically in 1973 for the first time by direct binding studies of mammalian brain and intestines with the plant alkaloid morphine (PERT & SNYDER 1973). Ligands that were found to bind to the opioid receptor were called henceforward opioids. Extensive studies with agonists and antagonists of opioid receptors revealed that numerous opioids showed distinct profiles of pharmacological activity suggesting that opioids activate three major groups of opioid receptor types: δ-, κ- and μ-receptors (reviewed in MARTIN 1979). Cloning of the three genes corresponding to the G-protein coupled receptors δ, κ and μ revealed a 50-60% sequence identity among these three genes differing mainly at the N- and C-termini (EVANS et al. 1992, CHEN et al. 1993a and CHEN et al. 1993b). Additionally, the non-classical orphan nociceptin receptor ORL (opioid receptor like) was identified as a fourth novel member of the opioid receptor family based on its high degree of structural homology towards the classical types (MOLLEREAU et al. 1994). Simultaneously with the discovery of opioid receptors the idea evolved that the original function of these receptors was not to mediate pharmacological effects of opium alkaloids but rather physiological effects of endogenous ligands. The identification of two smaller pentapeptides, Met- and Leu-enkephalin, in pig brain (HUGHES et al. 1975) and the isolation of the peptide β-endorphin from camel pituitary glands (LI & CHUNG 1976) was soon followed by the discovery of numerous other endogenous opioid peptides (reviewed in CORBETT et al. 2006). The majority of these opioid peptides varied in their affinity towards the δ, κ and μ receptor. Pharmacological effects of the non-peptide opioid morphine were instead believed to be mediated exclusively *via* the μ binding site (KOSTERLITZ 1985). The development of a μ-receptor knock out mouse confirmed that mainly responsible for the analgetic effect of morphine was indeed its binding to the μ-receptor (SORA et al. 1997). It could be shown later on, that very close morphine biosynthetic precursors also revealed a strong biological activity on the μ-opioid receptor while more distant biosynthetic precursors were assumed to not play a significant role in μ-opioid receptor signalling based on low efficacies and potencies (NIKOLAEV et al. 2007). Whereas it was accepted that endorphine and enkephaline that were isolated from brain are endogenous natural ligands of the opioid receptors, it was generally considered controversial that the plant alkaloid morphine could be an endogenous opioid, too.

The hypothesis that the opium poppy alkaloid could be endogenously produced in animals was based on the detection of morphine in mammalian tissue of subjects that did not receive any morphine (Tab. 1). More importantly, it was shown on various occasions that morphine is a naturally excreted metabolite in urine of rats (DONNERER et al. 1987) as well as humans (MIKUS et al. 1994, HOFMANN et al. 1999). The presence of the alkaloid in urine of humans was recently clearly verified by the lab of M. Spiteller, Dortmund, Germany, by demonstrating with an elegant morphine derivatization method the urinary excretion of the µ-receptor ligand in more than 20 tested subjects that were not administered any opioid (LAMSHÖFT& SPITELLER 2009, in preparation).

Against the hypothesis of an endogenous origin were reports on the presence of the alkaloid in food as well as in human and cow milk (HAZUM et al. 1981, ROWELL et al. 1982). Moreover, dubious and unreproducible results risked scientific sincerity. In several examples found in literature severe experimental error or contamination cannot be excluded. For instance, a novel morphine specific hypothetical receptor, µ3, was described, which was shown to be activated by morphine and morphine related precursors but not by endogenous opioid peptides (STEFANO et al. 1993, STEFANO et al. 1995, CADET et al. 2003). However, biological activity could not be observed in experiments with opioids and the cloned hypothetical µ3-receptor that were aimed to reproduce these results (unpublished, personal communication V. NIKOLAEV, University of Würzburg, Germany). Studies with µ3-receptor expressing HEK cell membranes and radioactive labeled [^{3}H]-naloxone, a µ-opioid receptor competitive antagonist exhibiting an extremely high affinity, did not reveal any binding of the radioligand to the hypothetical µ3-receptor suggesting that the receptor is inactive with opioids (unpublished, personal communication V. NIKOLAEV). Stefano and co-workers further attempted to show that human white blood cells synthesize morphine from tyramine *via* conversion to dopamine by human cytochrome P450 enzyme CYP 2D6 (ZHU et al. 2005) but these findings could not be reproduced by BOETTCHER et al. (2006) suggesting that human white blood cells fail to synthesize morphine. While some researchers in the field were challenged despite sensitive analytical methods with morphine concentrations below the limit of detection, some quantities of the alkaloid that were claimed to be of endogenous origin seemed very high, e.g. 368 pmol morphine/g FW rat adrenal gland (GOUMON & STEFANO 2000), 372 pmol morphine/g FW human heart tissue (ZHU et al. 2001b) or 4098 pmol morphine/g FW *Ascaris suum* (GOUMON et al. 2000c) (Tab. 1). Furthermore, GOUMON et al. (2000b) observed 35 times more morphine in the rat brain than GUARNA et al. (1998). These ambiguous results needed clarification and an approach that excluded laboratory contamination.

How can one distinguish between endogenously produced and exogenously introduced alkaloid? The only proof for a biosynthetic origin of morphine in mammals was derived from feeding studies with heavy-isotope labeled precursors. For the first time, the evidence for an endogenous synthesis of morphine in human cell lines was brought by M.H. Zenk's group in Halle, Germany, by conducting experiments with $^{18}O_2$ and later on precursor feeding studies with ^{13}C- and ^2H-labeled morphine precursors (POEAKNAPO et al. 2004, BOETTCHER et al. 2005). With a change of the analytical method from the previously used methods for the detection of morphine by RIA, ELISA or CE-HPLC to GC-MS a clear and specific determination of position-specific incorporation of heavy-isotope label was made possible. But even though MS is a sensitive, selective and reliable analytical method some challenges still remained and were addressed in this thesis. Criticism was raised that force feeding to cells outside of the human body could give ambiguous results. The hypothesis that a living animal is capable of biosynthesizing morphine from distant precursors was proven for the first time in this thesis with *in vivo* experiments on mice. Evidence for single biosynthetic steps within the morphine pathway was additionally brought with the discovery of new enzymatic functions of known mammalian enzymes.

Within this thesis, experiments were conducted with ^3H-labeled morphine to determine the stability of the alkaloid under different extraction and storage conditions. It could be clearly shown that 100% of morphine was recovered after the SPE-procedure with Strata X-C columns (Phenomenex) both in absence and presence of animal tissue such as rat and mouse brain. Within these studies a surprising stability of morphine over seven days in up to 2 % ammonium hydroxide in methanol was found. Morphine was also stable after evaporation and reconstitution in different solvents as well as after hydrolysis at 110°C for 40 min in 2 N HCl. Polypropylene tubes were found to be best for the work-up and drying procedures while glass tubes retained the alkaloid. Interestingly, some additives to the LC-solvent such as ammonium acetate, ammonium formate, acetic acid, formic acid and ammonium hydroxide were found to have severe effects on signal intensity resulting in differences of almost ten times depending on the additive (Fig. 8). However, with a 4000 QTRAP LC-MS instrument (Applied Biosystems) that showed for morphine originally a very promising low limit of detection of 0.1 pmol (31 pg), a severe ion suppression of the MS instrument was noted after the sample work-up through SPE. This effect could not be diminished and a 100% recovery after the SPE work-up, as it was determined for the radio-labeled morphine by scintillation counting and radio-TLC, could not be reached by LC-MS for the stable-isotope labeled morphine. Even after an intensive LC-method optimization, spiked morphine into an SPE-treated sample still revealed the same level of ion suppression that was assumed to be caused by compounds eluting with morphine from the SPE column. A change to a second MS instrument, an LTQ-Orbitrap (Thermo Scientific), at our collaborating institute in Dortmund, Germany, and to a new 100% validated SPE procedure with Bond Elut Certify (Varian) revealed a recovery of 85-90% morphine from body fluids. However, a recovery of only 50% and 20% for samples purified with the new Bond Elut Certify SPE method in absence or presence of biological background, respectively, was achieved with the 4000 QTRAP system. These results suggested that the ion source and possibly ion trapping in the MS instrument play a tremendous role for the analysis of morphine in animal tissue and body fluids. Alternatively, derivatization of a sample containing morphine, a standard procedure for the GC-MS analysis of morphine, could selectively improve the analyte's purity, remove undesired background interference and thus increase recoveries. The group in Dortmund for instance decreased the limit of detection for morphine recovered from human urine up to ten times by adding one derivatization step after the SPE work-up (LAMSHÖFT & SPITELLER 2009, in preparation).

Because of our experience of ion suppression with the 4000 QTRAP MS instrument, the ultrasensitive and highly selective LTQ-Orbitrap MS instrument was the ultimate choice for the analysis of the urinary excretion of morphine precursors and morphine by mice i.p. injected with distant biosynthetic precursors. Within this thesis the biosynthesis in living animals was unequivocally demonstrated in three parts. The first i.p. injection into mice of norlaudanosoline, a distant morphine precursor that has been found in animal brain and urine multiple times (Tab. 6), led to the urinary excretion of salutaridine, a morphinan alkaloid that is the precursor in the opium poppy morphine pathway. A following i.p. injection of salutaridine into mice resulted in the formation of salutaridinol and thebaine that were detected as urinary metabolites. The endogenous formation of thebaine, the first alkaloid with the pentacyclic morphine skeleton, was further verified by i.p. injection of [7D]-salutaridinol and incorporation of the deuterium label into thebaine. In a third part the existence of the bifurcate pathway in mammals was verified by i.p. injection of [N-CD$_3$]-thebaine that led to the urinary excretion of [N-CD$_3$]-codeine, [N-CD$_3$]-oripavine and, for the first time, [N-CD$_3$]-morphine. The i.p. injection of heavy-isotope labeled precursors such as [N-CD$_3$]-(R)-reticuline, [7D]-salutaridinol and [N-CD$_3$]-thebaine and detection of heavy-isotope labeled urinary metabolites proved for the first time that these biosynthetic steps are endogenously catalyzed in the living animal. If mammals are capable of biosynthesizing morphine from distant precursors what are the enzymatic catalysts that are involved in these biotransformations? Within this thesis enzymatic studies were conducted that discovered enzymes involved in selected biosynthetic steps of mammalian morphine biosynthesis. These enzyme studies strongly supported the results of the *in vivo* experiments and led within this thesis to the proposal of a biosynthetic pathway of morphine in animals which is depicted in Fig.42 and discussed in the following.

It was unequivocally shown herein that living animals are capable of biosynthesizing salutaridine from (*R,S*)-norlaudanosoline. This first part in the biosynthesis of morphine in animals is poorly investigated on the enzymatic level. The important morphinan alkaloid salutaridine escaped detection previously in application experiments of norlaudanosoline on rat brain as well as in incubations of rat liver with norlaudanosoline and reticuline (MEYERSON *et al.* 1979, CASHAW *et al.* 1983, KAMETANI *et al.* 1977, 1980). Since we clearly demonstrated the occurrence of salutaridine in urine of mice i.p. injected with norlaudanosoline the question was raised as to what types of reactions and catalysts are underlying the biotransformation of norlaudanosoline to salutaridine. It was proposed by BOETTCHER *et al.* (2005) that the biosynthesis of morphine in humans differed in its initial steps from plant morphine biosynthesis but is identical in its terminal reactions. For the initial steps in the biosynthesis of morphine in humans a sequence of methylation reactions was proposed in POEAKNAPO (2005) but also acknowledged that this sequence as well as underlying enzymes and stereochemistry required further investigations. This predicted sequence of reaction was exactly as it is postulated in this thesis in Fig. 42. Norlaudanosoline was proposed to be first *O*-methylated on the hydroxy group at either C4′ or C6 yielding 4′-*O*-methylnorlaudanosoline or 6-*O*-methylnorlaudanosoline following a 6- or 4′-*O*-methylation to form norreticuline, respectively. A third methylation on the ring nitrogen forms reticuline, the substrate for the phenol-coupling reaction yielding salutaridine. However, it was shown in this thesis that the i.p. injection of not only (*R,S*)- but also (*R*)-norlaudanosoline resulted in the urinary excretion of salutaridine. It was reported by BOETTCHER *et al.* (2005) on the contrary that (*R*)-norlaudanosoline was not incorporated into morphine when fed to SH-SY5Y cells suggesting that (*S*)-norlaudanosoline is the first alkaloid in morphine biosynthesis. These findings supported reports on the occurrence of the (*S*)-enantiomer of norlaudanosoline in rat and human brain (HABER *et al.* 1997, SANGO *et al.* 2000). Even with our very sensitive analytical method we were not able to detect morphine in urine of mice i.p. injected with (*R*)-norlaudanosoline but we found the morphine precursor salutaridine. The more distant positioned intermediate in the biosynthesis of morphine was unequivocally verified as a urinary metabolite. The discrepancy in results of the two studies could be

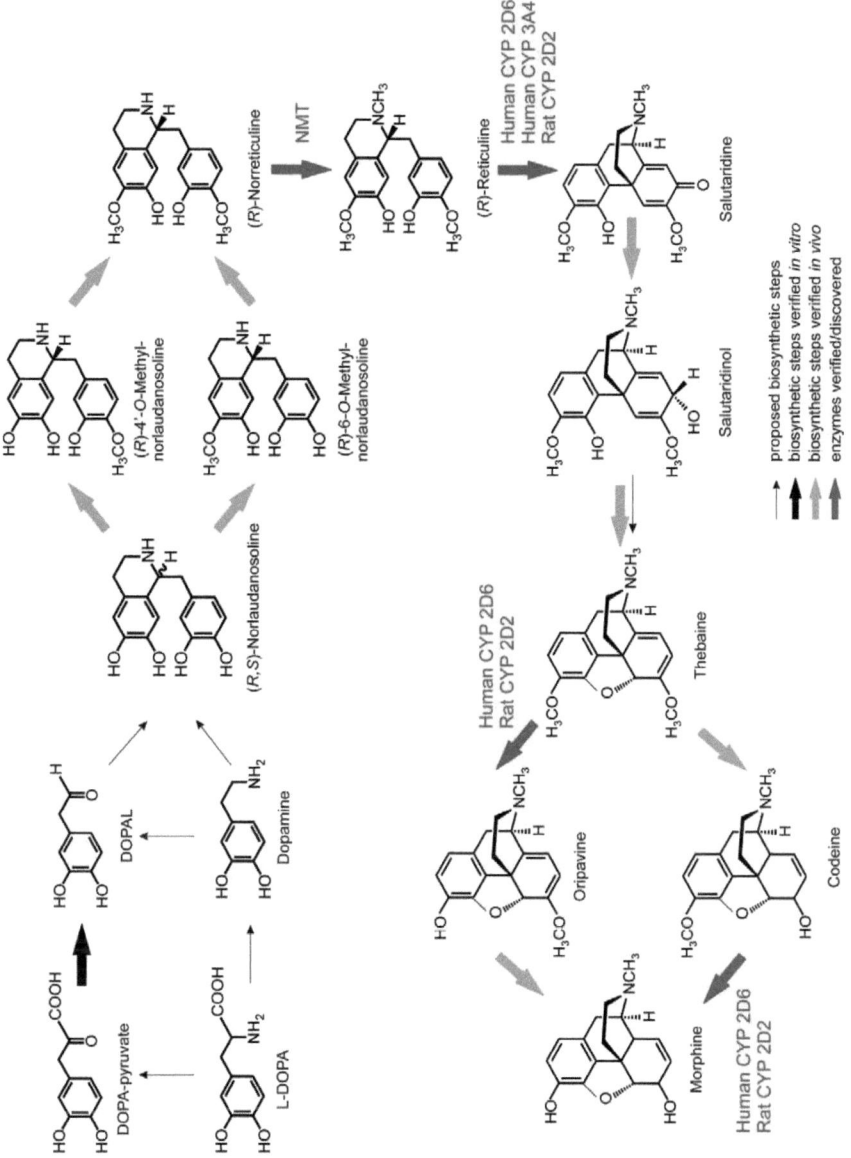

Fig. 42: Discoveries of biosynthetic steps (bold arrows) described in this thesis, contributing to the elucidation of the entire pathway of morphine in mammals.

explained by the fact that we analyzed in a HR-MS full scan modus the underivatized sample that enabled us to also look for more distant precursors of morphine biosynthesis. This is different from the method used by BOETTCHER et al. (2005) in which with *N*-methyl-*N*-TMS-trifluoroacetamide derivatized samples were subjected to GC-MS/MS preventing the simultaneous detection of more distant precursors. Based on the findings described in this thesis it cannot be confirmed yet that the (*S*)- rather than (*R*)-configurated THBIQ alkaloid could be the first precursor in morphine biosynthesis in mammals. Future studies on this Pictet-Spengler type reaction might reveal if there is a stereospecific enzyme involved in this condensation reaction that could possibly lead to a rise in the concentration of one of the two enantiomers of norlaudanosoline. This is for instance true for the condensing enzyme in opium poppy, norcoclaurine synthase, that enantio-selectively synthesizes the (*S*)-configurated alkaloid precursor in the opium poppy pathway (RUEFFER et al. 1981, STADLER et al. 1989, SAMANANI et al. 2004, MINAMI et al. 2007). An enzymatically catalyzed synthesis was also suggested in animals for the (*R*)-enantiomer of salsolinol, a TIQ alkaloid isolated from various human brain areas (MUSSHOFF et al. 2000). Chiral analysis of incubation mixtures of rat liver protein with DOPAL and dopamine as well as with heavy-isotope labeled L-DOPA and dopamine revealed the formation of equal amounts of the two enantiomers of norlaudanosoline indicating a spontaneous non-enzymatically catalyzed reaction (III.C.2). Taking our findings together the proposed initial steps in the morphine pathway of mammals as shown in Fig. 42 is favored in that the condensation of dopamine and DOPAL could also non-enzymatically yield racemic norlaudanosoline of which only the (*R*)-enantiomer will be further channelled into the biosynthesis of morphine.

One enzyme that could be involved in the morphine pathway in animals yielding salutaridine from norlaudanosoline is catechol-*O*-methyltransferase which catalyzes the 6-*O*-methylation of THBIQ alkaloids such as norlaudanosoline (COLLINS et al. 1973) without stereospecificity as it was shown for norcoclaurine-6-*O*-methyltransferases isolated from different plants (RUEFFER et al. 1983, SATO et al. 1994, FRICK & KUTCHAN 1999, OUNAROON et al. 2003). The second methyltransferase that is required for the formation of salutaridine from norlaudanosoline is 4′-*O*-methyltransferase which has not been found yet in animals. That mammals are capable of catalyzing a 4′-*O*-methylation of THBIQ alkaloids was strongly indicated by the occurrence of 4′-*O*-methylnorlaudanosoline in DAN-G cells (POEAKNAPO 2005) as well as by the urinary excretion of the THPB, morphinan and aporphine alkaloids after i.p. injection of (*R,S*)- and (*R*)-norlaudanosoline (III.B.1.1 and III.B.1.2). It is assumed that the 4′-*O*-methylating enzyme has not been detected yet because of low activities that disappear in incubations with for instance crude liver homogenates suggesting that the purified enzyme is needed to detect these minute amounts of product formation. The third enzyme that is required in the biotransformation of salutaridine from norlaudanosoline is an *N*-methyltransferase. In this thesis two human *N*-methyltransferases that could possibly catalyze the reaction were investigated, phenylethanolamine *N*-methyltransferase (EC 2.1.1.28) and a second enzyme named amine *N*-methyltransferase (EC 2.1.1.49). Only amine *N*-methyltransferase was active with THBIQ alkaloids as substrate and revealed a new interesting characteristic by catalyzing the *N*-methylation of exclusively (*R*)-configurated THBIQ alkaloids. This absolute stereospecificity has never been observed before for this enzyme and revealed at the same time a fundamental difference to the opium poppy biosynthetic pathway since in plants the *N*-methylation of coclaurine to *N*-methylcoclaurine is not (*R*)- but (*S*)-specific (LOEFFLER *et al.* 1995). This new enzymatic activity is additional evidence for the biosynthesis of morphine in animals. Incubations of a bovine isoform of this *N*-methyltransferase with TIQ alkaloids revealed a loose stereospecificity of the bovine enzyme (BAHNMAIER *et al.* 1999). The preference for the (*R*)- or (*S*)-configurated substrate switched in these studies depending on the degree of methylation at the isoquinoline ring. Within this thesis it was shown for the first time that the human *N*-methyltransferase accepted four (*R*)-configurated THBIQ alkaloids as substrate all of which represent potential precursors of morphine biosynthesis. This *N*-methyltransferase had previously been shown to convert tryptamine as its best substrate into *N*-methyltryptamine and was therefore called indolethylamine *N*-methyltransferase (THOMPSON & WEINSHILBOUM 1998, THOMPSON *et al.* 1999). However, a kinetic characterization for human *N*-methyltransferase described in this thesis (III.C.3.3) clearly demonstrated that (*R*)-norreticuline was the better substrate for the recombinant protein, although a K_m of 900 µM could still be quite high considering the *in vivo* situation. The finding of (*R*)-norreticuline as the best substrate suggested that as postulated by BOETTCHER *et al.* (2005) in mammalian morphine biosynthesis the two *O*-methylation steps at C6 and C4′ occur before the *N*-methylation. The methylation of (*R*)-norreticuline to (*R*)-reticuline could therefore be the final methylation step before the phenol-coupling reaction. In opium poppy, on the contrary, (*S*)-reticuline is produced from (*S*)-4′-hydroxy-*N*-methylcoclaurine *via* 4′-*O*-methyltransferase (FRENZEL & ZENK 1990). In plants, (*S*)-reticuline then undergoes subsequently a change of configuration to form (*R*)-reticuline, the ultimate precursor of the *C*-*C*-phenol-coupling reaction yielding salutaridine. The same inversion of configuration from (*S*)-reticuline to (*R*)-reticuline *via* formation of the quaternary 1,2-dehydroreticulinium ion was proposed by BOETTCHER *et al.* (2005) for the biosynthesis of morphine in humans. Attempts to feed heavy-isotope labeled 1,2-dehydroreticuline to SH-SY5Y cells failed probably due to the inability of the quaternary alkaloid to permeate the cell membrane. Taking these findings together with the results of this thesis it can be postulated that if (S)-norlaudanosoline is the first precursor in mammalian morphine biosynthesis as suggested by BOETTCHER et al. (2005) an inversion of configuration into (R) must occur before the N-methylation probably at the 1,2-dehydronorreticuline level. An enzyme known to form iminium ions from tertiary amines is the flavin-dependent monoamine oxidase (CHIBA et al. 1984). It could be possible that one of the two isoforms of monoamine oxidase A or B might be involved in the inversion of the configuration from (S) into (R) during morphine biosynthesis in animals which needs to be further investigated.

As it was described in this thesis, the i.p. injection of norlaudanosoline (III.B.1.1 and III.B.1.2) and (R)-[N-CD$_3$]-reticuline (III.B.1.3) into mice resulted also in the urinary excretion of THPB, aporphine and morphinan alkaloids. Formation of THPB after administration of norlaudanosoline or reticuline to rats or after incubation of reticuline with rat liver microsomes had been reported previously numerous times (CASHAW et al. 1974, 1983, MEYERSON et al. 1979, KAMETANI et al. 1972, 1977, 1980). Based on experiments with radio-labeled [^{14}C-methyl]-SAM or heavy-isotope labeled substrate it was postulated that the N-methyl group of reticuline is incorporated into tetrahydroprotoberberines in an enzymatically catalyzed conversion (CASHAW et al. 1974, KAMETANI et al. 1977, 1980) as it was shown earlier for the plant berberine bridge enzyme (BARTON et al. 1963b, BATTERSBY et al. 1963b). Whether or not an enzyme is involved in the formation of THPB alkaloids in animals from THBIQ alkaloids needs to be further investigated.

The formation of the aporphine and morphinan alkaloids corytuberine, isoboldine, pallidine and salutaridine after administration of norlaudanosoline and (R)-[N-CD$_3$]-reticuline described in this thesis indicated that the phenol-coupling reaction takes place in the living animal endogenously. Two contradictory results were reported from KAMETANI et al. (1977, 1980) who had not detected salutaridine in incubations with rat liver microsomes and WEITZ et al. (1987) who had claimed to have found in vivo as well as in vitro evidence for the formation of the phenol-coupled product salutaridine in animals. A first confirmation that mammalian enzymes are capable of catalyzing the phenol-coupling reaction came with the discovery of a P450 enzyme purified from porcine liver (AMANN et al. 1995). Both AMANN (1994) and FISINGER (1998) detected four products after incubation of mammalian liver microsomes with reticuline as substrate in the presence of an NADPH-generating system but only FISINGER (1998) suggested next to salutaridine the formation of isoboldine as a second out of four products. The identification of the three mammalian enzymes, human CYP 2D6, human CYP 3A4 and rat CYP 2D2, that were all found to catalyze the phenol coupling reaction in animals as described in detail in this thesis (III.C.3) involved also the elucidation of the four metabolites produced in that reaction. The formation of the four phenol-coupled products salutaridine, isoboldine, pallidine and corytuberine was determined by MS and was found to be true for all three mammalian P450 enzymes. These phenol-coupling catalyzing mammalian P450 enzymes resembled opium poppy salutaridine synthase in their requirement for NADPH and O_2 (AMANN 1994). While plant salutaridine synthase showed a temperature optimum of 20-25°C (GERARDY & ZENK 1993b), the mammalian P450 enzyme catalyzed the reaction best at 37°C (AMANN 1994). With the analysis of the characteristics of the three phenol-coupling reaction catalyzing P450 enzymes human CYP 2D6, human CYP 3A4 and rat CYP 2D2, it became clear that the strict substrate- and stereospecificity of the plant enzyme did not apply for the mammalian enzymes. The promiscuity of the mammalian P450 enzymes could be attributed to the sheer number of P450 catalysts in the animal and plant kingdom: 272 P450 enzymes are found in the plant genome of *Arabidopsis thaliana* while only 57 P450 enzymes are present in humans. An amino acid sequence comparison between the three mammalian enzymes catalyzing the formation of salutaridine from (*R*)-reticuline in morphine biosynthesis with salutaridine synthase from opium poppy revealed a very low homology of 16-18% (Tab. 15) whereas primary sequence motifs typical for eukaryotic P450 enzymes have been observed in all four sequences. These sequence motifs were for instance a conserved proline-rich region close to the N-terminus of the protein (YAMAZAKI *et al.* 1993), an I-helix with a conserved threonine that might be possibly important for oxygen activation (POULOS *et al.* 1987, LI & POULOS 2004) and a heme-binding region containing a cysteine required for iron binding (YOSHIOKA *et al.* 2001). Based on the low homology between plant and animal enzymes catalyzing the same reaction in morphine biosynthesis it seems as if during evolution two separate enzyme systems evolved independently. The crystal structures of human CYP 3A4 and human CYP 2D6 are known (WILLIAMS *et al.* 2004, ROWLAND *et al.* 2006) and plant salutaridine synthase was just recently cloned from opium poppy (GESELL *et al.* 2009), however, information about the three dimensional structure of the plant P450 enzyme is not yet available. Knowledge about substrate binding sites could further help to interpret why the mammalian P450 enzymes, human CYP 3A4, human CYP 2D6 and rat CYP 2D2, recognize and metabolize multiple and diverse chemical structures whereas salutaridine synthase from opium poppy is highly substrate-specific.

Tab. 15: Percentage sequence identities of salutaridine synthase with human CYP 2D6, human CYP 3A4 and rat CYP 2D2 aligned by "ClustalW".

	Salutaridine synthase (opium poppy)	human CYP 2D6	human CYP 3A4
human CYP 2D6	18		
human CYP 3A4	16	19	
rat CYP 2D2	18	71	19

The reaction mechanism of the phenol-coupling of (R)-reticuline to salutaridine is identical in plants and animals and is assumed to proceed through a single cycle of P450 action as depicted in Fig. 43 rather than a diradical mechanism via radical pairing as originally proposed by BARTON & COHEN (1957). The proposal of the mechanism described herein avoids the energetic instability of formal diradical species. The reaction starts with a fully activated P450 enzyme ("compound I" form) that abstracts a hydrogen from one phenolic hydroxy group of (R)-reticuline to create a phenoxy radical. The substrate rotates then in juxtaposition to the P450 iron ("compound II" form) that again abstracts a hydrogen from the second phenolic hydroxy group of (R)-reticuline. That leads to immediate coupling yielding a water molecule and salutaridine due to an enolization of the cyclohexanedienal ring.

Without any experimental evidence, it had been claimed that human CYP 2D6 might catalyze the transformation of (R)-reticuline to salutaridine (ZHU et al. 2005). In a recent paper, HAWKINS & SMOLKE (2008) accepted that claim and introduced CYP 2D6 into a recombinant yeast strain that produced racemic (R,S)-reticuline from commercially available (R,S)-norlaudanosoline. Feeding of the newly engineered yeast strain with (R,S)-norlaudanosoline led seemingly to the formation of a new product that was designated "salutaridine" in an average yield of 7%. Identification of "salutaridine" by LC-MS/MS was not based on direct experimental evidence but rather on comparison with expected MS data reported in literature for salutaridine (RAITH et al. 2003). The authors did not realize that CYP 2D6 catalyzes the phenol-coupling reaction of (R)- and (S)-reticuline resulting in eight different structures in the recombinant yeast strain, each alkaloid in different but consistent concentration. The yield of the correct and desired morphine precursor was therefore not 7% but rather less than 0.2%.

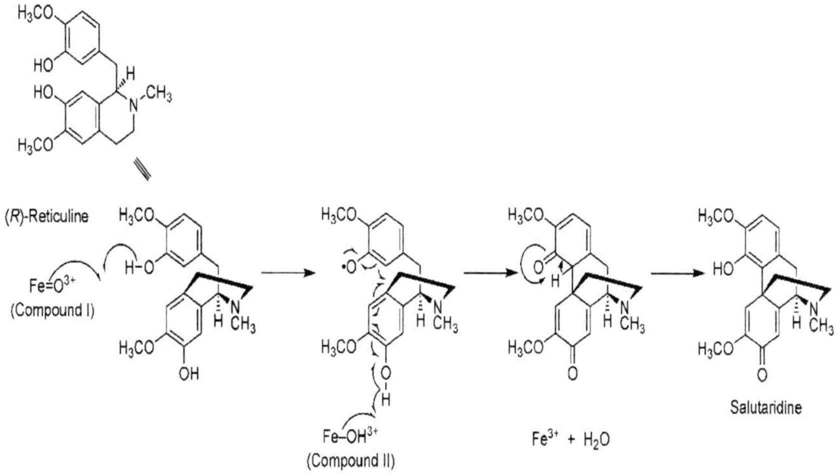

Fig. 43: Proposed mechanism of the oxidative phenol coupling reaction in mammals. The formation of salutaridine from (*R*)-reticuline is catalyzed by CYP 2D6 and CYP 3A4 and passes through a single cycle of iron oxidation.

The formation of thebaine after i.p. injection of salutaridine and [7D]-salutaridinol into mice confirmed unequivocally that living animals are capable of catalyzing the closure of the oxide bridge to form the ether linkage in the pentacyclic structure of morphine alkaloids (III.B.2). This biotransformation has been described earlier in *in vitro* studies with rats (FISINGER 1998) and later shown with precursor feeding experiments on human neuroblastoma cells (BOETTCHER *et al.* 2005). How do animals catalyze this reaction of oxide bridge closure that chemically follows an S_N2'-elimination? As postulated in Fig. 42 it is assumed that in animals, like in plants, salutaridine is first reduced to salutaridinol in a stereospecific, enzymatically catalyzed reaction. An indication that the reduction reaction occurred stereospecifically in an NADPH-dependent manner was found by incubation experiments of cytosolic rat liver protein with salutaridine yielding salutaridinol but not the diastereomer *epi*-salutaridinol (FISINGER 1998). The finding was also verified in this thesis describing that i.p. injection of salutaridine led to the urinary excretion of salutaridinol whereas *epi*-salutaridinol was not detected (III.B.2.1). Additionally, salutaridinol and not the epimer was shown to be a precursor of human morphine biosynthesis in human neuroblastoma cells (BOETTCHER *et al.* 2005). That was also confirmed within this thesis for the living animal as heavy-isotope labeled thebaine was formed only from [7D]-salutaridinol and not from the heavy-isotope labeled epimer (III.B.2.2). In plants the reduction of salutaridine to salutaridinol catalyzed by salutaridine reductase (GERARDY & ZENK 1993a, ZIEGLER et al. 2006) is followed by an ester activation to salutaridinol-7-O-acetate catalyzed by the CoA dependent enzyme salutaridinol-7-O-acetyltransferase (LENZ & ZENK 1995b, GROTHE et al. 2001) yielding thebaine either spontaneously or enzymatically catalyzed. In vitro studies with acetyl CoA and rat liver protein reported the formation of thebaine from salutaridinol (FISINGER 1998). Therefore, it seems likely that such an ester formation could also be possible in the morphine pathway in mammals which will need to be further investigated in future studies.

Within this thesis the presence of the terminal steps of morphine biosynthesis as shown in Fig. 42 were clearly demonstrated when [N-CD$_3$]-morphine was shown to be unequivocally formed in the living animal from [N-CD$_3$]-thebaine proving an endogenous biosynthesis. The simultaneous occurrence of [N-CD$_3$]-oripavine and [N-CD$_3$]-codeine after i.p. injection of [N-CD$_3$]-thebaine was an evidence for the bifurcate pathway. An additional i.p. injection of oripavine led to the urinary excretion of morphine and verified for the first time that living animals are capable of catalyzing the vinyl-ether cleavage of oripavine to morphine in the bifurcate pathway. Enzymatic studies revealed that next to human CYP 2D6 and human CYP 3A4, which both had long been known to be involved in dealkylation reactions of xenobiotics, also rat CYP 2D2 was found to be a promiscuous enzyme. Whereas human CYP 3A4 N-demethylated codeine to N-norcodeine (CARACO et al. 1996), it was shown that only the two P450 enzymes of the 2D family were capable to catalyze the 3-O-demethylation of thebaine and codeine in the bifurcate pathway of mammalian morphine biosynthesis. Taking that into account, human CYP 2D6 and rat CYP 2D2 are unique P450 enzymes in that each catalyzes three reactions in mammalian morphine biosynthesis assigning a possibly important role for these biocatalysts in the pathway. The usual reaction mechanism commonly accepted for the demethylation reaction catalyzed by P450 enzymes is shown in Fig. 44 and is initiated with a hydrogen atom abstraction from the methoxy group at C3 of codeine or thebaine by the iron of the fully activated P450 ("compound I" form). That is followed by a hydroxylation of the so formed methoxy group carbon radical ("compound II" form) yielding formaldehyde and the demethylated product. With codeine or thebaine as substrate the incubation with human CYP 2D6 and rat CYP 2D2 yielded the 3-O-demethylated products morphine or oripavine, respectively. It is interesting to note that as in plants, both alternative routes towards morphine starting from thebaine co-exist but it still remains unclear why these two pathways evolved and are used equally in opium poppy to produce morphine.

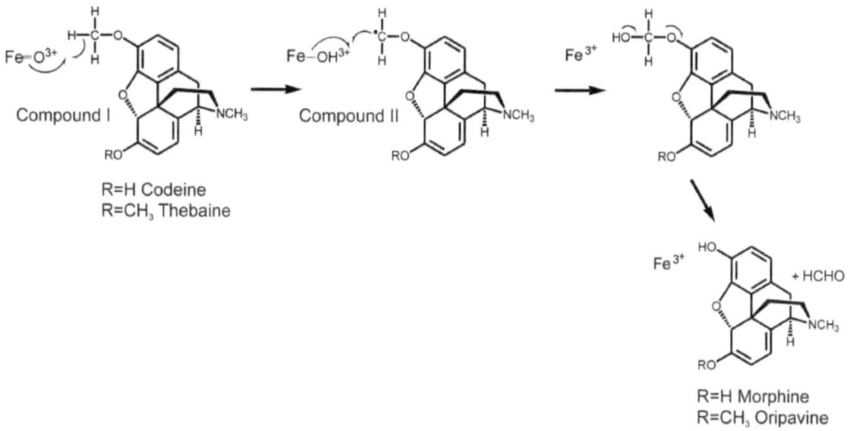

Fig. 44: Mechanism for the 3-*O*-demethylation of codeine and thebaine catalyzed by human CYP 2D6 and rat CYP 2D2 modified from GUENGERICH *et al.* (1996).

Are the amounts detected in animals physiologically relevant? The urinary excretion of morphine precursors and morphine after i.p. injection of distant precursors of morphine biosynthesis into mice were unequivocally shown in this thesis and were found to be of low concentration. Reason for the minute amounts of urinary excreted morphine could be that the produced morphine precursor and morphine itself disappear in the metabolism of the animal after i.p. injection. Assuming that some alkaloid conjugates could have also escaped detection and that the metabolic fate of i.p. injected substance into mice remains unknown we essentially only took a snapshot of the metabolism by analyzing the urine of i.p. injected mice. In all application studies the recovery of i.p. injected substance was low and decreased even more with the i.p. injection of less distant precursors of the pathway's terminal steps. While 20-40% (*R,S*)- and (*R*)-norlaudanosoline were recovered as urinary metabolite only 16% of i.p. injected (*R*)-[*N*-CD$_3$]-reticuline were found. Recoveries decreased even more to 0.3% of i.p. injected salutaridine and 1.2% of i.p. injected [7D]-salutaridinol, possibly partly due to the hydrolysis step that was skipped, to the lowest recoveries of 0.3% of i.p. injected oripavine and 0.01% of i.p. injected [*N*-CD$_3$]-thebaine. The metabolic fate of these biosynthetic morphine precursors is unknown. The question whether the minute amounts of morphine detected in urine of mice i.p. injected with morphine precursors (e.g. i.p. injection of [*N*-CD$_3$]-thebaine yielded 1.7 nM [*N*-CD$_3$]-morphine detected in urine) or that are found in animal tissue (Tab. 1) could be physiologically relevant can be addressed by the findings of NIKOLAEV *et al.* (2007). Based on µ-receptor mediated G$_i$-protein activation studies in living cells it was reported that the *ca.* 10 nM concentration of morphine found in neuroblastoma cells (BOETTCHER *et al.* 2005) and mouse cerebellum (MULLER *et al.* 2008) could be biologically active at the µ-receptor. Morphine bound to proteins as well as morphine conjugates such as sulfate esters or glucuronides represent morphine pools that could add up all together to a much higher amount of morphine being present in mammalian tissue and body fluids than it has been previously estimated or detected in this thesis. Morphine-6-sulfate and morphine-6-glucuronide, both conjugates that have been considered an *in vivo* detoxification mechanism for morphine, have been shown to have more analgesic potency than morphine itself (MORI *et al.* 1972, PAUL *et al.* 1989, CROOKS *et al.* 2006). Taken together with the finding that the analgesic effect of these conjugates was drastically decreased in µ-receptor knockout mice (LOH *et al.* 1998) it seems reasonable that morphine conjugates could be the actual effective form of morphine analgesia. It is also entirely conceivable that in certain physiological or pathological conditions endogenous morphine concentrations could be increased and thus could relieve pain by possibly stimulating the µ-receptor. In an early report by DONNERER *et al.* (1987), demonstration of that hypothesis was attempted for arthritic rats which showed an up to three times elevated urinary excretion of morphine compared to control rats. Reliable analytical techniques are now available (HOFMANN *et al.* 1999, LAMSHÖFT & SPITELLER 2009, in preparation) to determine highly selectively by MS minute amounts of morphine in tissue or body fluids. These techniques can be used to further expand the studies by DONNERER *et al.* (1987) to elucidate whether animals including humans suffering for instance chronic or acute pain show elevated endogenous morphine levels.

While plants produce morphine as a secondary metabolite to protect against predators and to thus increase survivability, the role of endogenous morphine in animals is not yet resolved. A possible function of morphine as a neurotransmitter was supported by BIANCHI *et al.* (1993). They observed by immunocytochemical approaches that [^{3}H]-morphine is accumulated in neurons. It was further described that endogenous morphine was released *via* a Ca^{2+}-dependent mechanism from rat brain slices in response to depolarization after exposure to high potassium concentrations (GUARNA *et al.* 1998). Recently, MULLER *et al.* (2008) aimed to show that endogenous morphine fulfills criteria that define a neurotransmitter, i.e. showing the substance's presence in neurons, its release and the presence of specific receptors on the postsynaptic cell. The localization of morphine was indeed reported for specific brain regions of the mouse CNS, specifically in nerve termini of Purkinje cells that had previously been shown to express also the µ-opioid receptor (MRKUSICH *et al.* 2004). Additionally, MULLER *et al.* (2008) observed that morphine is secreted *via* a Ca^{2+}-dependent mechanism into the culture medium after stimulation of human neuroblastoma cells (SH-SY5Y) with nicotine, a well-established model for studying secretion of neurotransmitters, and that the cellular response to morphine occurred at concentrations as low as 0.1 nM. Interestingly, human CYP 2D6 mRNA and protein as well as human CYP 3A4 mRNA, both P450 enzymes that have been shown to catalyze the C-C-phenol-coupling of (R)-reticuline to salutaridine, were found to be present in specific brain regions (SIEGLE et al. 2001, AGARWAL et al. 2008) suggesting that the site of morphine biosynthesis could be close to the site of action in the central nervous system. The elucidation of gene products that are involved in the biosynthesis in animals could be a helpful tool and could lead to the discovery of interesting pharmacological targets thus adding credibility for a physiological role of endogenous morphine. With this thesis progress has been made in that direction by showing that living animals are capable of biosynthesizing morphine and by identifying enzymes that are possibly involved in selected biosynthetic steps of the morphine pathway. A physiological role of endogenous morphine could be elucidated with gene-specific knockout mice. The mouse homolog of CYP 2D6 has not been identified yet and could be a gene target, even though a similar genetic situation as in humans possessing two non-related P450 enzymes that are able to catalyze the phenol-coupling reaction could also be conceivable for mice. The mouse homolog of the stereospecific NMT on the contrary seems to be a better gene target since this enzyme might play a unique and crucial role in morphine biosynthesis in animals. Medicinal studies of the knockout mice compared to control mice including analysis of the receptor response, such as pain perception experiments, as well as monitoring phaenotypical changes, such as mood and behaviour, could deliver important observations for the understanding of the functional role of morphine genes as well as the pharmacology and physiology of endogenous morphine itself. If it is shown that endogenous morphine is involved in pain modulation and brain physiology in animals and that its synthesis is dependent on enzymes that are postulated to be involved in the pathway new pharmacological targets could be designed. These promising perspectives will unify a network of research in biochemistry, chemistry, molecular biology and medicine and require that the scientific community is open to the idea that mammals are capable of synthesizing morphine endogenously which has been shown unequivocally in living animals within this thesis.

V Summary

The controversy whether morphine in mammals is of exogenous or endogenous origin was investigated by analyzing urinary metabolites of mice i.p. injected with potential precursors of the morphine pathway. In addition, enzymes of selected biosynthetic steps were characterized.

1) The stability of morphine under various extraction and storage conditions was determined. Conditions were chosen that resulted in a 100% recovery of radio-labeled morphine (spec. act. 80 Ci/mmol) extracted from a biological matrix such as rat or mouse brain tissue.

2) However, severe ion suppression was observed with the 4000 Triple Quadrupole Mass Spectrometry instrument leading to a 60-80% loss of signal when stable-isotope labeled morphine extracted by the same method as above in the absence or presence of biological matrix. These shortcomings were circumvented by use of the Linear Trap Quadrupole Orbitrap High-Resolution Mass Spectrometer resulting in an 80-95% recovery of morphine extracted from biological matrices by solid phase extraction (Bond Elut Certify).

3) High-Resolution Mass Spectrometry with an Orbitrap Mass Spectrometer identified metabolites of the morphine pathway in urine of mice i.p. injected with distant, biosynthetic morphine precursors. The tetrahydrobenzylisoquinoline alkaloid norlaudanosoline has previously been detected multiple times in animals and humans. I.p. injection of a mammalian key intermediate, norlaudanosoline, led to the urinary excretion of salutaridine, a central intermediate also in plant morphine biosynthesis. In addition, i.p. injection of (R)-$[N$-$C^2H_3]$-reticuline yielded $[N$-$C^2H_3]$-salutaridine as a urinary metabolite.

4) Application of salutaridine resulted in the formation of two alkaloids salutaridinol and thebaine which were detected in the urine of i.p. injected mice. This demonstrates that the living animal is capable of catalyzing the closure of the oxide bridge in the pentacyclic morphine skeleton analogous to the corresponding opium poppy enzyme. Additionally, $[7$-$^2H]$-salutaridinol and its epimer $[7$-$^2H]$-*epi*-salutaridinol were chemically synthesized and separately i.p. injected into mice. The diastereomer $[7$-$^2H]$-salutaridinol and not $[7$-$^2H]$-*epi*-salutaridinol, which is biologically inactive also in the plant, was converted to $[7$-$^2H]$-thebaine demonstrating the enzyme-catalyzed reaction.

5) I.p. injection of [N-C^{2}H$_3$]-thebaine into mice led to the urinary excretion of [N-C^{2}H$_3$]-codeine, [N-C^{2}H$_3$]-oripavine and [N-C^{2}H$_3$]-morphine confirming that in animals, as in plants, a bifurcate pathway evolved yielding morphine from thebaine *via* two alternative routes. These two parallel routes in the bifurcate pathway start from thebaine that is converted to morphine either *via* codeine or *via* oripavine (Fig. 42). Oripavine injection yielded morphine demonstrating that animals are capable of catalyzing the long sought cleavage reaction of the vinyl-ether of oripavine.

6) The quantitative analysis of the metabolites from the *in vivo* studies revealed severe losses of i.p. injected substances recovered in the urine of the mice (norlaudanosoline: 60-80%, (*R*)-[N-C^{2}H$_3$]-reticuline: 84%, salutaridine: 99%, [7-^{2}H]-salutaridinol: 99%, [N-C^{2}H$_3$]-thebaine: 99%, oripavine: 99%). The inability to recover most of the morphine pathway intermediates demonstrates that the majority of these physiologically active compounds is either degraded or metabolically inactivated (glucuronidation, sulfation, acetylation, *etc.*).

7) 3,4-Dihydroxyphenylacetaldehyde is formed from dopamine *via* oxidative deamination catalyzed by monoamine oxidase. Liquid Chromatography Mass Spectrometry analysis also identified 3,4-dihydroxyphenylacetaldehyde in incubations of 3,4-dihydroxyphenyl-pyruvate with mouse liver protein providing new evidence for an alternative formation of the aldehyde. The existence of this second route was additionally shown by incubation of [ring-^{13}C$_6$]-L-DOPA with unlabeled dopamine leading to the formation of (*R*,*S*)-[^{13}C$_6$]-norlaudanosoline carrying the heavy-isotope label in the benzyl moiety.

8) A new enzymatic function of a known human enzyme was discovered catalyzing the *N*-methylation exclusively of (*R*)-configured morphine precursors. Kinetic analysis revealed that (*R*)-norreticuline was the best substrate for this human *N*-methyltransferase (K_m = 900 µM, k_{cat} / K_m = 0.056 mM^{-1} s^{-1}) yielding (*R*)-reticuline, the precursor of the phenol-coupling reaction in morphine biosynthesis.

9) The key reaction in morphine biosynthesis, the *C*-*C*-phenol-coupling of (*R*)-reticuline to salutaridine, is catalyzed by human CYP 2D6 (K_m = 2.5 µM, k_{cat} / K_m = 0.08 mM^{-1} s^{-1}), human CYP 3A4 (K_m = 993 µM, k_{cat} / K_m = 0.04 mM^{-1} s^{-1}) and rat CYP 2D2 (K_m = 0.7 µM, k_{cat} / K_m = 53.9 mM^{-1} s^{-1}). Addition of cytochrome b_5 increased the catalytical efficiency of CYP 3A4 approximately three-fold. Whereas plant salutaridine synthase only forms salutaridine as a product, all three mammalian enzymes catalyzed the formation of a set of four products from (*R*)-reticuline: the *para-ortho* coupled product

salutaridine, the *ortho-ortho* coupled product (-)-corytuberine, the *para-para* coupled product (+)-pallidine and the *ortho-para* coupled product (-)-isoboldine. These four phenol-coupled products were also detected in urine of mice i.p. injected with (*R*)-[*N*-C^{2}H$_3$]-reticuline.

10) Rat CYP 2D2 and not, as published before, rat CYP 2D1 was found to catalyze the 3-*O*-demethylation of thebaine to oripavine and of codeine to morphine, two reactions that occur in the bifurcate pathway of morphine biosynthesis. Kinetic parameters and catalytic efficiencies were determined for human CYP 2D6 catalyzing the 3-*O*-demethylation of thebaine to oripavine (K_m = 48 µM, k_{cat} / K_m = 1.6 mM^{-1} s^{-1}) and rat CYP 2D2 catalyzing the 3-*O*-demethylation of thebaine to oripavine (K_m = 42 µM, k_{cat} / K_m = 25 mM^{-1} s^{-1}) and codeine to morphine (K_m = 29 µM, k_{cat} / K_m = 8 mM^{-1} s^{-1}).

Based on the results of this thesis the pathway of morphine in mammals has been largely elucidated.

VI References

ADLER, T.K., FUJIMOTO, J.M., WAY, E.L. and BAKER, E.M. (1955). The metabolic fate of codeine in man. *J. Pharmacol. Exp. Ther.* **114**: 251-262.

AGARWAL, V., KOMMADDI, R.P., VALLI, K., RYDER, D., HYDE, T.M., KLEINMAN, J.E., STROBEL, H.W. and RAVINDRANATH, V. (2008). Drug metabolism in human brain: high levels of cytochrome P4503A43 in brain and metabolism of anti-anxiety drug alprazolam to its active metabolite. *PloS One* **3**: e2337.

AMANN, T. (1994). Untersuchungen zur enzymatischen Umsetzung von (*R*)-Retikulin zu Salutaridin, dem Schlüsselschritt der Morphinbiosynthese im Säugetier. *Dissertation*. Ludwig-Maximilians-Universität München.

AMANN, T. and ZENK, M.H. (1991). Formation of the morphine precursor salutaridine is catalyzed by a cytochrome P-450 enzyme in mammalian liver. *Tetrahedron Lett.* **32**: 3675-3678.

AMANN, T., ROOS, P.H., HUH, H. and ZENK, M.H. (1995). Purification and characterization of a cytochrome P450 enzyme from pig liver, catalyzing the phenol oxidative coupling of (*R*)-reticuline to salutaridine, the critical step in morphine biosynthesis. *Heterocycles* **40**: 425-440.

ANSHER, S.S. and JAKOBY, W.B. (1986). Amine *N*-methyltransferase from rabbit liver. *J. Biol. Chem.* **261**: 3996-4001.

ARAKAWA, Y., IMAI, K. and TAMURA, Z. (1979). High-performance liquid chromatographic determination of dopamine sulfoconjugates in urine after L-DOPA administration. *J. Chromatogr.* **162**: 311-318.

AXELROD, J. (1961). Enzymatic formation of psychomimetic metabolites from normally occurring compounds. *Science* **134**: 343.

AXELROD, J. (1962). Purification and properties of phenylethanolamine *N*-methyltransferase. *J. Biol. Chem.* **237**: 1657-1660.

AXELROD, J. and INSCOE, J.K. (1960). Glucuronide formation of narcotic drugs *in vitro* and *in vivo*. *Proc. Soc. Exp. Biol. Med.* **103**: 675-676.

BAHNMAIER, A.H., WOESLE, B. and THOMAS, H. (1999). Stereospecific *N*-methylation of the tetrahydroisoquinoline alkaloids isosalsoline and salsolidine by amine *N*-methyltransferase A from bovine liver. *Chirality* **11**: 160-165.

BARNES, R.A. (1964). The structure of salutaridine. *An. Acad. Brasil. Cienc.* **36**: 238-239.

BARTON, D.H.R. and COHEN, T. (1957). Some biogenetic aspects of phenol oxidation. Festschrift Arthur Stoll. Birkhäuser Verlag, Basel.

BARTON, D.H.R., BHAKUNI, D.S., JAMES, R. and KIRBY, G.W. (1967). Phenol oxidation and biosynthesis. Part XII. Stereochemical studies related to the biosynthesis of the morphine alkaloids. *J. Chem. Soc. (C)*: 128-132.

BARTON, D.H.R., HESSE, R.H. and KIRBY, G.W. (1963b). The origin of the "berberine carbon". *Proc. Chem. Soc.*: 267.

BARTON, D.H.R., KIRBY, G.W., STEGLICH, W. and THOMAS, G.M. (1963a). The biosynthesis and synthesis of morphine alkaloids. *Proc. Chem. Soc.*: 203-204.

BARTON, D.H.R., KIRBY, G.W., STEGLICH, W., THOMAS, G.M., BATTERSBY, A.R., DOBSON, T.A. and RAMUZ, H. (1965). Investigations on the biosynthesis of morphine alkaloids. *J. Chem. Soc.* **65**: 2423-2428.

BATTERSBY, A.R. (1963). Tilden lecture. The biosynthesis of alkaloids. *Proc. Chem. Soc.* 189-200.

BATTERSBY, A.R. and BINKS, R. (1960). Biosynthesis of morphine: Formation of morphine from norlaudanosoline. *Proc. Chem. Soc.*: 360-361.

BATTERSBY, A.R., BINKS, R., FOULKES, D., MCIS, R.J., MCCALDIN, D.J. and RAMUZ, H. (1963a). 1-Benzylisoquinoline as precursors of the opium alkaloids: tracer and stereochemical studies. *Proc. Chem. Soc.*: 203.

BATTERSBY, A.R., BINKS, R., FRANCIS, R.J., MCCALDIN, D.J. and RAMUZ, H. (1964). Alkaloid biosynthesis. Part IV. 1-Benzylisoquinolines as precursors of thebaine, codeine, and morphine. *J. Chem. Soc.*: 3600-3610.

BATTERSBY, A.R., FOULKES, D.M. and (in part) BINKS, R. (1965). Alkaloid biosynthesis. Part VIII. Use of optically active precursors for investigations on the biosynthesis of morphine alkaloids. *J. Chem. Soc.*: 3323-3332.

BATTERSBY, A.R., FRANCIS, R.J., HIRST, M. and STAUNTON, J. (1963b). Biosynthesis of the "berberine bridge". *Proc. Chem. Soc.*: 268.

BATTERSBY, A.R., JONES, R.C.F. and KAZLAUSKAS, R. (1975). Experiments on the early steps of morphine biosynthesis. *Tetrahedron Lett.* **22+23**: 1873-1876.

BIANCHI, E., ALESSANDRINI, C., GUARNA, M. and TAGLIAMONTE, A. (1993). Endogenous codeine and morphine are stored in specific brain neurons. *Brain Res.* **627**: 210-215.

BLUME, A.J., SHORR, J., FINBERG, J.P. and SPECTOR, S. (1977). Binding of the endogenous nonpeptide morphine-like compound to opiate receptors. *Proc. Natl. Acad. Sci. USA* **74**: 4927-4931.

BOETTCHER, C., FELLERMEIER, M., BOETTCHER, C., DRÄGER, B. and ZENK, M.H. (2005). How human neuroblastoma cells make morphine. *Proc. Natl. Acad. Sci. USA* **102**: 8495-8500.

BOETTCHER, C., FISCHER, W. and ZENK, M.H. (2006). Comment on "Human white blood cells synthesize morphine: CYP 2D6 modulation." *J. Immunol.* **176**: 5703-5704.

BRADFORD, P.J. (1976). A rapid and sensitive method for the quantitation of microgram quantities of protein utilizing the principle of protein-dye binding. *Anal. Biochemistry* **72**: 248-254.

BREESE, G.R., CHASE, T.N. and KOPIN, I.J. (1969). Metabolism of some phenylethylamines and their β-hydroxylated analogs in brain. *J. Pharmacol. Exp. Ther.* **165**: 9-13.

BROCHMANN-HANSSEN, E. (1984). A second pathway for the terminal steps in the biosynthesis of morphine. *Planta Med.* **50**: 343-345.

BROCHMANN-HANSSEN, E. (1985). Biosynthesis of morphinan alkaloids. In: J.D. Phillipson, M.F. Roberts and M.H. Zenk, editors, *The chemistry and biology of isoquinoline alkaloids*. Springer Berlin, 229-239.

CADET, P., MANTIONE, K.J. and STEFANO, G.B. (2003). Molecular identification and functional expression of mu 3, a novel alternatively spliced variant of the human mu opiate receptor gene. *J. Immunol.* **170**: 5118-5123.

CARACO, Y., TATEISHI, T., GUENGERICH, F.P. and WOOD, A.J.J. (1996). Microsomal codeine *N*-demethylation: cosegregation with cytochrome P4503A4 activity. *Drug Metab. Dispos.* **24**: 761-764.

CASHAW, J.L. (1993a). Tetrahydropapaveroline in brain regions of rats after acute ethanol administration. *Alcohol.* **10**: 133-138.

CASHAW, J.L. (1993b). Determination of tetrahydropapaveroline in the urine of Parkinsonian patients receiving L-DOPA-carbidopa (Sinemet) therapy by high-performance liquid chromatography. *J. Chromatogr.* **613**: 267-273.

CASHAW, J.L., GERAGHTY, C.A., MCLAUGHLIN, B. and DAVIS, V.E. (1987). A method for determination of subpicomole concentrations of tetrahydropapaveroline in rat brain by high-performance liquid chromatography with electrochemical detection. *Anal. Biochem.* **162**: 274-282.

CASHAW, J.L., MCMURTREY, K.D., BROWN, H. and DAVIS, V. (1974). Identification of catecholamine-derived alkaloids in mammals by gas chromatography and mass spectrometry. *J. Chromatogr.* **99**: 567-573.

CASHAW, J.L., RUCHIRAWAT, S., NIMIT, Y. and DAVIS, V.E. (1983). Regioselective *O*-methylation of tetrahydropapaveroline and tetrahydroxyberbine *in vivo* in rat brain. *Biochem. Pharmacol.* **32**: 3163-3169.

CHEN, Y., MESTEK, A., LIU, J. and YU, L. (1993a). Molecular cloning of a rat κ opioid receptor reveals sequence similarities to the μ and κ opioid receptors. *Biochem. J.* **295**: 625-628.

CHEN, Y., MESTEK, A., LIU, J., HURLEY, J.A. and YU, L. (1993b). Molecular cloning and functional expression of a mu-opioid receptor from rat brain. *Mol. Pharmacol.* **44**: 8-12.

CHEN, Z.R., SOMOGYI, A.A. and BOCHNER, F. (1988). Polymorphic *O*-demethylation of codeine. *Lancet* **332**: 914-915.

CHIBA, K., TREVOR, A. and CASTAGNOLI, JR., A. (1984). Metabolism and the neurotoxic tertiary amine, MPTP, by brain monoamine oxidase. *Biochem. Biophys. Res. Comm.* **120**: 574-578.

COLLINS, A.C., CASHAW, J.L. and DAVIS, V.E. (1973). Dopamine-derived tetrahydroisoquinoline alkaloids-inhibitors of neuramine metabolism. *Biochem. Pharmacol.* **22**: 2337-2348.

COLLINS, M.A. (2004). Tetrahydropapaveroline in Parkinson's disease and alcoholism: A look back in honor of Merton Sandler. *Neurotoxicology* **25**: 117-120.

CONNETT, R.J and KIRSHNER, N. (1970). Purification and properties of bovine phenylethanolamine *N*-methyltransferase. *J. Biol. Chem.* **245**: 329-334.

CORBETT, A.D., HENDERSON, G., MCKNIGHT, A.T. and PATERSON, S.J. (2006). 75 years of opioid research: the exciting but vain quest for the holy grail. *Brit. J. Pharmacol.* **147**: S153-S162.

CORDELL, G.A. (1981). Introduction to alkaloids. A biogenetic approach. J. Wiley & Sons. New York.

CROOKS, P.A., KOTTAYIL, S.G., AL-GHANANEEM, A.M., BYRN, S.R. and BUTTERFIELD, D.A. (2006). Opiate receptor binding properties of morphine-, dihydromorphine- and codeine 6-*O*-sulfate ester congeners. *Bioorg. Med. Chem Lett.* **16**: 4291-4295.

DAVIS, V.E. and WALSH, M.J. (1970). Alcohol, amines and alkaloids: A possible biochemical basis for alchohol addiction. *Science* **167**: 1005-1008.

DAYER, P., DESMEULES, J., LEEMANN, T. and STRIBERNI, R. (1988). Bioactivation of the narcotic drug codeine in human liver is mediated by the polymorphic monooxygenase catalyzing debrisoquine 4-hydroxylation (cytochrome P-450 db/bufI). *Biochem. Biophys. Res. Commun.* **152**: 411-416.

DONNERER, J., CARDINALE, G., COFFEY, J., LISEK, C.A., JARDINE, I. and SPECTOR, S. (1987). Chemical characterization and regulation of endogenous morphine and codeine in the rat. *J. Pharmacol. Exp. Ther.* **242**: 583-587.

DONNERER, J., OKA, K., BROSSI, A., RICE, K.C. and SPECTOR, S. (1986). Presence and formation of codeine and morphine in the rat. *Proc. Natl. Acad. Sci. USA* **83**: 4566-4567.

DUNCAN, R.J.S. (1975). The preparation of some biochemically important aldehydes. *Can. J. Biochem.* **53**: 920-922.

EICHELBAUM, M. and GROSS, A.S. (1990). The genetic polymorphism of debrisoquine/sparteine metabolism-clinical aspects. *Pharmac. Ther.* **46**: 377-394.

EICHELBAUM, M., SPANNBRUCKER, N., STEINCKE, B. and DENGLER, H.J. (1979). Defective *N*-oxidation of sparteine in man: a new pharmacological defect. *Eur. J. Clin. Pharmacol.* **16**: 183-187.

EICHELBAUM, R.M., SPANNBRUCKER, N., STEINCKE, B. and DENGLER, H.J. (1979). Defective *N*-oxidation of sparteine in man: a new pharmacogenetic defect. *Eur. J. Clin. Pharmacol.* **16**: 183-187.

EVANS, C.J., KEITH, JR., D.E., MORRISON, H., MAGENDZO, K., and EDWARDS, R.H. (1992). Cloning of delta opioid receptor by functional expression. *Science* **258**: 1952-1955.

FISH, F. and HAYES, T.S. (1974). Hydrolysis of morphine glucuronide. *J. Forensic Sci.* **19**: 676-683.

FISINGER, U. (1998). Untersuchungen zur Morphinbiosynthese in der Ratte *Rattus rattus* L. und im Schlafmohn *Papaver somniferum* L. *Dissertation*. Ludwig-Maximilians-Universität München.

FRENZEL, T. and ZENK, M.H. (1990). *S*-Adenosyl-L-methionine-(*S*)-coclaurine-4'-*O*-methyltransferase, a regio- and stereoselective enzyme of the (*S*)-reticuline pathway. *Phytochemistry* **29**: 3505-3011.

FRICK, S. and KUTCHAN, T.M. (1999). Molecular cloning and functional expression of *O*-methyltransferases common to isoquinoline alkaloid and phenylpropanoid biosynthesis. *Plant J.* **17**: 329-339.

GATES, M. and TSCHUDI, G. (1952). The synthesis of morphine. *J. Am. Chem. Soc.* **74**: 1109-1110.

GATES, M. and TSCHUDI, G. (1956). The synthesis of morphine. *J. Am. Chem. Soc.* **78**: 1380-1393.

GERAGHTY, C.A. and CASHAW, J.L. (1989). Method for the identification of tetrahydropapaveroline using Pictet-Spengler condensation reaction and high-performance liquid chromatography. *J Chromatogr.* **489**: 399-403.

GERARDY, R. and ZENK, M.H. (1993a). Purification and characterization of salutaridine: NADPH 7-oxidoreductase from *Papaver somniferum*. *Phytochemistry* **34**: 125-132.

GERARDY, R. and ZENK, M.H. (1993b). Formation of salutaridine from (*R*)-reticuline by a membrane-bound cytochrome P-450 enzyme from *Papaver somniferum*. *Phytochemistry* **32**: 79-86.

GESELL, A., ROLF, M., ZIEGLER, J., DÍAZ-CHÁVEZ, M.L., HUANG, F.-C. and KUTCHAN, T.M. (2009). CYP719B1 is salutaridine synthase. the *C-C* phenol-coupling enzyme of morphine biosynthesis in opium poppy. *J. Biol. Chem.* **284**: 24432-24442.

GINTZLER, A.R., GERSHON, M.D. and SPECTOR, S. (1978). A non-peptide morphine-like compoundo: immunocytochemical localization in the mouse brain. *Science* **199**: 447-448.

GINTZLER, A.R., LEVY, A. and SPECTOR, S. (1976a). Antibodies as a means of isolating and characterizing biologically active substances: Presence of a non-peptide, morphine-like compound in the central nervous system. *Proc. Natl. Acad. Sci. USA* **73**: 2132-2136.

GINTZLER, A.R., MOHACSI, E. AND SPECTOR, S. (1976b). Radioimmunoassay for the simultaneous determination of morphine and codeine. *Eur. J. Pharmacol.* **38**: 149-156.

GOLDSTEIN, A., BARRETT, R.W., JAMES, I.F., LOWNEY, L.I., WEITZ, C.J., KNIPMEYER, L.L. and RAPOPORT, H. (1985). Morphine and other opiates from beef brain and adrenal. *Proc. Natl. Acad. Sci. USA* **82**: 5203-5207.

GOLLWITZER, J., LENZ, R., HAMPP, N. and ZENK, M.H. (1993). The transformation of neopinone to codeinone in morphine biosynthesis proceeds non-enzymatically. *Tetrahedron Lett.* **34**: 5703-5706.

GOUMON, Y. and STEFANO, G.B. (2000). Identification of morphine in the rat adrenal gland. *Mol. Brain Res.* **77**: 267-269.

GOUMON, Y., BOURET, S., CASARES, F., ZHU, W., BEAUVILLAIN, J.-C. and STEFANO, G.B. (2000b). Lipopolysaccharide increases endogenous morphine levels in rat brain. *Neurosci. Lett.* **293**: 135-138.

GOUMON, Y., CASARES, F., PRYOR, S., FERGUSON, L., BROWNAWELL, B., CADET, P., RIALAS, C.M., WELTERS, I.D.M., SONETTI, D. and STEFANO, G.B. (2000c). *Ascaris suum*, an intestinal parasite, produces morphine. *J. Immunol.* **165**: 339-343.

GOUMON, Y., CASARES, F., ZHU, W. and STEFANO, G.B. (2001). The presence of morphine in ganglionic tissues of *Modiolus deminissus*: a highly sensitive method of quantitation for morphine and its derivatives. *Mol. Brain Res.* **86**: 184-188.

GOUMON, Y., ZHU, W., WEEKS, B.S., CASARES, F., CADET, P., BOUGAEVA, M., BROWNAWELL, B. and STEFANO, G.B. (2000a). Identification of morphine in the adrenal medullary chromaffin PC-12 cell line. *Mol. Brain Res.* **81**: 177-180.

GROTHE, T., LENZ, R. and KUTHCAN, T.M. (2001). Molecular characterization of the salutaridinol-7-*O*-acetyltransferase involved in morphine biosynthesis in opium poppy. *J. Biol. Chem.* **276**: 30717-30723.

GUARNA, M., BIANCHI, E., BARTOLINI, A., GHELARDINI, N., GALEOTTI, N., BRACCI, L., NERI, C., SONETTI, D. and STEFANO, G. (2002). Endogenous morphine modulates acute thermonociception in mice. *J. Neurochem.* **80**: 271-277.

GUARNA, M., NERI, C., PETRIOLI, F. and BIANCHI, E. (1998). Potassium-induced release of endogenous morphine from rat brain slices. *J. Neurochem.* **70**: 147-152.

GUENGERICH, F.P., YUN, C.-H. and MACDONALD, T.L. (1996). Evidence for a 1-electron oxidation mechanism in N-dealkylation of N,N-dialkylanilines by cytochrome P450 2B1. *J. Biol. Chem.* **271**: 27321-27329.

GUENGERICH, F.P., MILLER, G.P., HANNA, I.H., SATO, H. and MARTIN, M.V. (2002). Oxidation of methoxyphenethylamines by cytochrome P450 2D6. Analysis of rate-limiting steps. *J. Biol. Chem.* **277**: 33711-33719.

GULLAND, J.M. and ROBINSON, R. (1925). The constitution of codeine and thebaine. *Mem. Proc. Manchester Lit. Phil. Soc.* **69**: 79-86.

HABER, H., ROSKE, I., ROTTMANN, M., GEORGI, M. and MELZIG, M.F. (1997). Alcohol induced formation of morphine precursors in the striatum of rats. *Life Sci.* **60**: 79-89.

HANNA, I.H., KRAUSER, J.A., CAI, H., KIM, M.S. and GUENGERICH, F.P. (2001). Diversity in mechanisms of substrate oxidation by cytochrome P450 2D6. Lack of an allosteric role of NADPH-cytochrome P450 reductase in catalytic regioselectivity. *J. Biol. Chem.* **276**: 39553-39561.

HAWKINS, K.M. and SMOLKE, C.D. (2008). Production of benzylisoquinoline alkaloids in *Saccharomyces cerevisiae*. *Nat. Chem. Biol.* **4**: 564-573.

HAZUM, E., SABATKA, J.J., CHANG, K., BRENT, D.A., FINDLAY, J.W.A. and CUATRECASAS, P. (1981). Morphine in cow and human milk: could dietary morphine constitute a ligand for specific morphine (μ) receptors? *Science* **213**: 1010-1012.

HIROI, T., CHOW, T., IMAOKA, S. and FUNAE, Y. (2002). Catalytic specificity of CYP2D isoforms in rat and human. *Drug Metab. Dispos.* **30**: 970-976.

HOFMANN, U., SEEFRIED, S., SCHWEIZER, E., EBNER, T., MIKUS, G. and EICHELBAUM, M. (1999). Highly sensitive gas chromatographic-tandem mass spectrometric method for the determination of morphine and codeine in serum and urine in the femtomolar range. *J. Chromatogr. B* **727**: 81-88.

HOLTZ, P. and HEISE, R. (1938). Fermentativer Abbau von l-Dioxyphenylalanin (Dopa) durch Niere. *N-S Arch. Ex. Path. Ph.* **191**: 87-119.

HOLTZ, P., STOCK, K. and WESTERMANN, E. (1963). Über die Blutdruckwirkung des Dopamins. *N-S Arch. Ex. Path. Ph.* **246**: 133-146.

HOLTZ, P., STOCK, K. and WESTERMANN, E. (1964). Formation of tetrahydropapaveroline from dopamine *in vitro*. *Nature* **203**: 656-658.

HUGHES, J., SMITH, T.W., KOSTERLITZ, H.W., FOTHERGILL, L.A., MORGAN, B.A. and MORRIS, H.R. (1975). Identification of two related pentapeptides from the brain with potent opiate agonist activity. *Nature* **258**: 577-579.

ICHINOSE, H., KUROSAWA, Y., TITANI, K., FUJITA, K. and NAGATSU, T. (1989). Isolation and characterization of a cDNA clone encoding human aromatic L-amino acid decarboxylase. *Biochem. Biophys. Res. Comm.* **164**: 1024-1030.

ISHIDA, T., ZANO, M. and TOKI, S. (1991). In vivo formation of codeinone and morphinone from codeine. *Drug Metab. Dispos.* **19**: 895-899.

JAFFE, J.H. and MARTIN, W.R. (1990). Opioid analgesics and antagonists. In: A. Goodman Gilman, T.W. Rall, A.S. Nies and P. Taylor, editors, *The Pharmacological Basis of Therapeutics*. Pergamon Press. New York. 485-521.

JEZ, J. and CAHOON, R.E. (2004). Kinetic mechanism of glutathionine synthetase from *Arabidopsis thaliana*. *J. Biol. Chem.* **279**: 42726-42731.

KAGAN, R.M. and CLARKE, S. (1994). Widespread occurrence of three sequence motifs in diverse S-adenosylmethionine-dependent methyltransferases suggests a common structure for these enzymes. *Arch. Biochem. Biophys.* **310**: 417-427.

KAMETANI, T. and FUKUMOTO, K. (1971). Synthesis of morphinanedienone by phenol oxidation and the Pschorr reaction. *Heterocycles* **8**: 341-356.

KAMETANI, T., IHARA, M. and TAKAHASHI, K. (1972). Biotransformation of (+)-reticuline in rat. *Chem. Pharm. Bull.* **20**: 1587-1588.

KAMETANI, T., KANAYA, N., OHTA, Y. and IHARA, M. (1980). Biotransformation of isoquinoline alkaloids with rat liver microsomes. *Heterocycles* **14**: 963-970.

KAMETANI, T., OHTA, Y., TAKEMURA, M., IHARA, M. and FUKUMOTO, K. (1977). Biotransformation of reticuline into corexamine, scoulerine, pallidine, and isoboldine with rat liver enzyme. *Bioorg. Chem.* **6**: 249-256.

KEITH, D.E., MURRAY, S.R., ZAKI, P.A., CHU, P.C., LISSIN, D.V., KANG, L., EVANS, C.J. and VON ZASTROW, M. (1996). Morphine activates opioid receptors without causing their rapid internalization. *J. Biol. Chem.* **271**: 19021-19024.

KILLIAN, A.K., SCHUSTER, C.R., HOUSE, J.T., SHOLL, S., CONNORS, M. and WAINER, B.H. (1981). A non-peptide morphine-like compound from brain. *Life Sci.* **28**: 811-817.

KNOX, W.E. and EDWARDS, S.W. (1955). Enzymes involved in conversion of tyrosine to acetoacetate. *Methods Enzymol.* **2**: 287-300.

KODAIRA, H. and SPECTOR, S. (1988). Transformation of thebaine to oripavine, codeine, and morphine by rat liver, kidney, and brain microsomes. *Proc. Natl. Acad. Sci. USA* **85**: 1267-1271.

KODAIRA, H., LISEK, C.A., JARDINE, J., ARIMURA, A. and SPECTOR, S. (1989). Identification of the convulsant opiate thebaine in mammalian brain. *Proc. Natl. Acad. Sci. USA* **86**: 716-719.

KOSTERLITZ, H.W. (1985). Has morphine a physiological function in the animal kingdom? *Nature* **317**: 671-672.

KUMAGAI, Y., TODAKA, T. and TOKI, S. (1990). A new metabolic pathway of morphine: *in vivo* and *in vitro* formation of morphinone and morphinone-glutathione adduct in guinea pig. *J. Pharmacol. Exp. Ther.* **255**: 504-510.

KUTCHAN, T.M. (1998). Molecular genetics of plant alkaloid biosynthesis. In: G. Cordell, editor, *The Alkaloids – Chemistry and Biology.* Academic Press. San Diego. **50**: 257-316.

LAIDLAW, P.P. (1910). The action of tetrahydropapaveroline hydrochloride. *J. Physiol.* **40**: 480-491.

LAMSHÖFT, M. and SPITELLER, M. (2009, in preparation). Highly sensitive liquid chromatography-triple quadrupole mass spectrometric method for the detection and quantitation of morphine in human urine. *J. Chomatogr. B.*

LENZ, R. (1994). Salutaridinol-7-*O*-Acetyltransferase und Codeinon-Reduktase: Zwei neue Enzyme in der Morphinbiosynthese. *Dissertation.* Ludwig-Maximilians-Universität München.

LENZ, R. and ZENK, M.H. (1995a). Purification and properties of codeinone reductase (NADPH) from *Papaver somniferum* cell cultures and differentiated plants. *Eur. J. Biochem.* **233**: 132-139.

LENZ, R. and ZENK, M.H. (1995b). Acetyl-coenzyme A: salutaridinol-7-*O*-acetyltransferase from *Papaver soniferum* plant cell cultures. The enzyme catalyzing the formation of thebaine in morphine biosynthesis. *J. Biol. Chem.* **270**: 31091-31096.

LI, C.H. and CHUNG, D. (1976). Isolation and structure of an untriakontapeptide with opiate activity from camel pituitary glands. *Proc. Nat. Acad. Sci. USA* **73**: 1145-1148.

LI, Y. and POULOS, T.L. (2004). Crystallization of cytochromes P450 and substrate-enzyme interactions. *Curr. Top. Med. Chem.* **4**: 1789-1802.

LIU, Y., BILFINGER, T.V. and STEFANO, G.B. (1997). A rapid and sensitive quantitation method of endogenous morphine in human plasma. *Life Sci.* **60**: 237-243.

LOEFFLER, S., DEUS-NEUMANN, B. and ZENK, M.H. (1995). *S*-Adenosyl-L-methionine-(*S*)-coclaurine-*N*-methyltransferase from *Tinospora cordifolia*. *Phytochemistry* **38**: 1387-1395.

LOH, H.H., LIU, H.-C., CAVALLI, A., YANG, W., CHEN, Y.F. and WEI, L.-N. (1998). μ Opioid receptor knockout in mice: Effects on ligand-induced analgesia and morphine lethality. *Mol. Brain Res.* **54**: 321-326.

MACKAY, G.D. and HODGKIN, D.C. (1955). A chrystallographic examination of the structure of morphine. *J. Chem. Soc.* 3261-3267.

MAHGOUB, A., IDLE, J.R., DRING, L.G., LANCASTER, R. and SMITH, R.L. (1977). Polymorphic hydroxylation of debrisoquine in man. *Lancet* **2**: 584-586.

MANNERING, G.J., DIXON, A.C., BAKER, E.M. and ASAMI, T. (1954). The *in vivo* liberation of morphine from codeine in man. *J. Pharmacol. Exp. Ther.* **111**: 142-146.

MARTIN, W.R. (1979). History and development of mixed opioid agonists, partial agonists and antagonists. *Br. J. Clin. Pharmac.* **7**: 273S-279S.

MATSUBARA, K., FUKUSHIMA, S., AKANE, A., KOBAYASHI, S. and SHIONO, H. (1992). Increased urinary morphine, codeine and tetrahydropapaveroline in Parkinsonian patient undergoing L-3,4-dihydroxyphenylalanine therapy: A possible biosynthetic pathway of morphine from 3,4-dihydroxyphenylalanine in humans. *J. Pharmacol. Exp. Ther.* **260**: 974-978.

MAVROJANNIS, M. (1903). L-Action Cataleptique de la Morphine chez les Rats. Contribution a la Theorie Toxique da la Catalepsie. *Comptes rendues Soc. Biol.* **55**: 1092-1094.

MEISSNER, W. (1819). Ueber ein neues Pflanzenalkali (Alkaloid). *Journal für Chemie und Physik.* **25**: 379-381.

MEYERSON, L.R., CASHAW, J.L., MCMURTREY, K.D. and DAVIS, V.E. (1979). Stereoselective enzymatic *O*-methylation of tetrahydropapaveroline and tetrahydroxyberbine alkaloids. *Biochem. Pharmacol.* **28**: 1745-1752.

MIKUS, G., BOCHNER, F., EICHELBAUM, M., HORAK, P., SOMOGYI, A.A. and SPECTOR, S. (1994). Endogenous codeine and morphine in poor and extensive metabolisers of the CYP 2D6 (Debrisoquine/sparteine) polymorphism. *J. Pharm. Exp. Ther.* **286**: 546-551.

MIKUS, G., SOMOGYI, A.A., BOCHNER, F. and EICHELBAUM, M. (1991a). Thebaine *O*-demethylation to oripavine: genetic differences between two rat strains. *Xenobiotica* **21**: 1501-1509.

MIKUS, G., SOMOGYI, A.A., BOCHNER, F. and EICHELBAUM, M. (1991b). Codeine *O*-demethylation: rat strain differences and the effects of inhibitors. *Biochem. Pharmacol.* **41**: 757-762.

MINAMI, H., DUBOUZET, E., IWASA, K. and SATO, F. (2007). Functional analysis of norcoclaurine synthase in *Coptis japonica*. *J. Biol. Chem.* **282**: 6274-6282.

MISRA, A.L., PONTANI, R.B. and MULÉ, S.J. (1974). Pharmacokinetics and metabolism of [^{3}H]thebaine. *Xenobiotica* **4**: 17-32.

MOLLEREAU, C., PARMENTIER, M., MAILLEUX, P., BUTOUR, J.-L., MOISAND, C., CHALON, P., CAPUT, D., VASSART, G. and MEUNIER, J.-C. (1994). ORL1, a novel member of the opioid receptor family. Cloning, functional expression and localization. *FEBS Lett.* **341**: 33-38.

MORI, M., OGURI, K., YOSHIMURA, H., SHOMOMURA, K., KAMATA, O. and UEKI, S. (1972). Chemical synthesis and analgesic effect of morphine ethereal sulfates. *Life Sci.* **11**: 525-533.

MRKUSICH, E.M., KIVELL, B.M., MILLER, J.H. and DAY, D.J. (2004). Abundant expression of mu and delta opioid receptor mRNA and protein in the cerebellum of the fetal, neonatal, and adult rat. *Dev. Brain Res.* **148**: 213-222.

MULLER, A., GLATTARD, E., TALEB, E., KEMMEL, V., LAUX, A., MIEHE, M., DELADANDE, F., ROUSEEL, G., DORSSELAER, A.V., METZ-BOUTIQUE, M.-H., AUNIS, D. and GOUMON, Y. (2008). Endogenous Morphine in SH-SY5Y cells and the mouse cerebellum. *PloS ONE* **3**: e1641.

MUSSHOFF, F., SCHMIDT, P., DETTMEYER, R., PRIEMER, F., JACHAU, K. and MADEA, B. (2000). Determination of dopamine and dopamine-derived (*R*)-/(*S*)-salsolinol and norsalsolinol in various human brain areas using solid-phase extraction and gas chromatography/ mass spectrometry. *Forensic Sci. Int.* **113**: 359-366.

MYERS, R.D. and MELCHIOR, C.L. (1977). Alcohol drinking: Abnormal drinking caused by tetrahydropapaveroline in brain. *Science* **196**: 554-556.

NERI, C., GHELARDINI, C., SOTAK, B., PALMITER, R.D., GUARNA, M., STEFANO, G.B. and BIANCHI, E. (2008). Dopamine is necessary to endogenous morphine formation in mammalian brain *in vivo*. *J. Neurochem.* **106**: 2337-2344.

NIKOLAEV, V.O., BOETTCHER, C., DEES, C., BÜNEMANN, M., LOHSE, M.J. and ZENK, M.H. (2007). Live-cell monitoring of µ-opioid receptor mediated G-protein activation reveals strong biological activity of close morphine biosynthetic precursors. *J. Biol. Chem.* **282**: 27126-27132.

OBERST, F.W. (1941). Relationship of the chemical structure of morphine derivatives to their urinary excretion in free and bound forms. *J. Pharmacol. Exp. Ther.* **73**: 401-404.

OKA, K., KANTROWITZ, J.D. and SPECTOR, S. (1985). Isolation of morphine from toad skin. *Proc. Natl. Acad. Sci. USA* **82**: 1852-1854.

OSCARSON, M., HIDESTRAND, M., JOHANSSON, I. and INGELMAN-SUNDBERG, M. (1997). A combination of mutations in the CYP 2D6*17 (CYP 2D6Z) allele causes alterations in enzyme function. *Mol. Pharmacol.* **52**: 1034-1040.

OUNAROON, A., DECKER, G., SCHMIDT, J., LOTTSPEICH, F. and KUTCHAN, T.M. (2003). (*R,S*)-Reticuline 7-*O*-methyltransferase and (*R,S*)-norcoclaurine 6-*O*-methyltransferase of *Papaver somniferum* – cDNA cloning and characterization of methyl transfer enzymes of alkaloid biosynthesis in opium poppy. *Plant J.* **36**: 808-819.

PAUL, D., STANDIFER, K.M., INTURRISI, C.E. and PASTERNAK, G.W. (1989). Pharmacological characterization of morphine-6-beta-glucuronide, a very potent morphine metabolite. *J. Pharmacol. Exp. Ther.* **251**: 477-483.

PERT, C.B. and SNYDER, S.H. (1973). Opiate receptor: Demonstration in nervous tissue. *Science* **179**: 1011-1014.

POEAKNAPO, C. (2005). Biosynthesis of endogenous morphine in human neuroblastoma SH-SY5Y cells. *Dissertation*. Martin-Luther-Universität Halle-Wittenberg.

POEAKNAPO, C., SCHMIDT, J., BRANDSCH, M., DRÄGER, B. and ZENK, M.H. (2004). Endogenous formation of morphine in human cells. *Proc. Natl. Acad. Sci. USA* **101**: 14091-14096.

POULOS, T.L., FINZEL, B.C. and HOWARD, A.J. (1987). High-resolution crystal structure of cytochrome P450$_{cam}$. *J. Mol. Biol.* **195**: 687-700.

PYMAN, F.L. (1909). CLXXXII.-Isoquinoline derivatives. Part II. The constitution of the reduction products of papaverine. *J. Chem. Soc., Trans.* **95**: 1610-1623.

RAITH, K., NEUBERT, R., POEAKNAPO, C., BOETTCHER, C., ZENK, M.H. and SCHMIDT, J. (2003). Electrospray tandem mass spectrometric investigations of morphinans. *J. Am. Soc. Mass. Spectrom.* **14**: 1262-1269.

RICE, K.C. and BROSSI, A. (1980). Expedient synthesis of racemic and optically active *N*-norreticuline and *N*-substituted and 6'-bromo-*N*-norreticulines. *J. Org. Chem.* **45**: 592-601.

ROBINSON, R. (1917). LXXV.-A theory of the mechanism of the phytochemical synthesis of certain alkaloids. *J. Chem. Soc., Trans.* **111**: 876-899.

ROWELL, F.J., SPARK, P. and RICH, C.G. (1982). Non-peptide morphine-like compounds in vertebrate brain tissue and food. *J. Pharmacol.* **77**: 461P.

ROWLAND, P., BLANEY, F.E., SMYTH, M.G., JONES, J.J., LEYDON, V.R., OXBROW, A.K., LEWIS, C.J., TENNANT, M.G., MODI, S., EGGLESTON, D.S., CHENERY, R.J. and BRIDGES, A.M. (2006). Crystal structure of human cytochrome P450 2D6. *J. Biol. Chem.* **281**: 7614-7622.

RUEFFER, M., EL-SHAGI, H., NAGAKURA, H. and ZENK, M.H. (1981). (*S*)-Norlaudanosoline synthase: The first enzyme in the benzylisoquinoline biosynthetic pathway. *FEBS Lett.* **129**: 5-9.

RUEFFER, M., NAGAKURA, H. and ZENK, M.H. (1983). Partial purification and properties of (*S*)-adenosyl-L-methionine: (*R*), (*S*)-norlaudanosoline-6-*O*-methyltransfersae from *Argemone platyceras* cell cultures. *Planta Med.* **49**: 131-137.

SÄLLSTRÖM BAUM, S., HILL, R., KIIANMAA, K. and ROMMELSPACHER, H. (1999). Effect of ethanol on (*R*)- and (*S*)-salsolinol, salsoline, and THP in the Nucleus Accumbens of AA and ANA rats. *Alcohol* **18**: 165-169.

SAMANANI, N., LISCOMBE, D.K. and FACCHINI. P.J. (2004). Molecular cloning and characterization of norcoclaurine synthase, an enzyme catalyzing the first committed step in benzylisoquinoline alkaloid biosynthesis. *Plant J.* **40**: 302-313.

SAMBROOK, J., FRITSCH, E.F. and MANIATIS, T. (1989). Molecular cloning: A laboratory manual. Cold Spring Harbor Laboratory. New York.

SANDLER, M., BONHAM CARTER, S., HUNTER, K.R. and STERN, G.M. (1973). Tetrahydroisoquinoline alkaloids: *in vivo* metabolites of L-DOPA in man. *Nature* **241**: 439-443.

SANGAR, M.C., ANANDATHEERTHAVARADA, H.K., TANG, W., PRABU, S.K., MARTIN, M.V., DOSTALEK, M., GUENGERICH, F.P. and AVADHANI, N.G. (2009). Human liver mitochondrial cytochrome P450 2D6 – individual variations and implications in drug metabolism. *FEBS J.* **276**: 3440-3453.

SANGO, K., MARUYAMA, W., MATSUBARA, K., DOSTERT, P., MINAMI, C., KAWAI, M. and NAOI, M. (2000). Enantio-selective occurrence of (*S*)-tetrahydropapaveroline in human brain. *Neruosci. Lett.* **283**: 224-226.

SATO, F., TSUJITA, T., KATAGIRI, Y., YOSHIDA, S. and YAMADA, Y. (1994). Purification and characterization of *S*-adenosyl-L-methionine:norcoclaurine 6-*O*-methyltransferase from cultured *Coptis japonica* cells. *Eur. J. Biochem.* **225**: 125-131.

SCHÖPF, C. (1927). Constitution of morphine alkaloids. *Ann.* **452**: 211-267.

SCHULZ, S., MAYER, D., PFEIFFER, M., STUMM, R., KOCH, T. and HÖLLT, V. (2004). Morphine induces terminal µ-opioid receptor desensitization by sustained phosphorylation of serine-375. *EMBO J.* **23**: 3282-3289.

SEEVERS, M.H., DAVIS, V.E. and WALS, M.J. (1970). Morphine and ethanol physical dependence: a critique of a hypothesis. *Science* **170**: 1113-1115.

SEIBERT, R.A., WILLIAMS, C.E. and HUGGINS, R.A. (1954). The isolation and identification of "bound" morphine. *Science* **120**: 222-223.

SERTÜRNER, F.W. (1806). Darstellung der reinen Mohnsäure (Opiumsäure) nebst einer chemischen Untersuchung des Opiums mit vorzüglicher Hinsicht auf einen darin neu entdeckten Stoff und die dahin gehörigen Bemerkungen. *J. Pharm. Ärzte Apotheker Chem.* **14**: 47-93.

SERTÜRNER, F.W. (1817). Über das Morphium, eine neue salzfähige Grundlage, und die Mekonsäure als Hauptbestandteil des Opiums. *Ann. Phys.* **25**: 56-90.

SHORR, J., FOLEY, K. and SPECTOR, S. (1978). Presence of a non-peptidemorphine-like compound in human cerebrospinal fluid. *Life Sci.* **23**: 2057-2062.

SIEGLE, I., FRITZ, P., ECKHARDT, K., ZANGER, U.M. and EICHELBAUM, M. (2001). Cellular localization and regional distribution of CYP 2D6 mRNA and protein expression in human brain. *Pharmacogenetics* **11**: 237-245.

SMITH, R.L. (1986). Introduction. *Xenobiotica* **16**: 361-365.

SORA, I., TAKAHASHI, N., FUNADA, M., UJIKE, H., REVAY, R.S., DONOVAN, D.M., MINER, L.L. and UHL, G.R. (1997). Opiate receptor knockout mice define µ receptor roles in endogenous nociceptive responses and morphine-induced analgesia. *Proc. Natl. Acad. Sci. USA* **94**: 1544-1549.

STADLER, R. and ZENK, M.H. (1990). A revision of the gernerally accepted pathway for the biosynthesis of the benzyltetrahydroisoquinoline allkaloid reticuline. *Liebigs Ann. Chem.*: 555-562.

STADLER, R., KUTCHAN, T.M. and ZENK, M.H. (1989). (S)-Norcoclaurine is the central intermediate in benzylisoquinoline alkaloid biosynthesis. *Phytochemistry* **28**: 1083-1086.

STEFANO, G.B., DIGENIS, A., SPECTOR, S., LEUNG, M.K., BILFINGER, T.V., MAKMAN, M.H., SCHARRER, B. and ABUMRAD, N.N. (1993). Opiate-like substances in an invertebrate, an opiate receptor on invertebrate and human immunocytes, and a role in immunosuppression. *Proc. Natl. Acad. Sci. USA* **90**: 11099-11103.

STEFANO, G.B., HARTMAN, A., BILFINGER, T.V., MAGAZINE, H.I., LIU, Y., CASARES, F. and GOLIGORSKY, M.S. (1995). Presence of the mu3-opiate receptor in endothelial cells. Coupling to nitric oxide production and vasodilation. *J. Biol. Chem.* **270**: 30290-30293.

STORK, G. (1960). The morphine alkaloids. In: R.H.F. Manske, editor, *The Alkaloids – Chemistry and Physiology*. Academic Press. New York. **6**: 219-245.

TABAKOFF, B., ANDERSON, R. and ALIVISATOS, S.G.A. (1973). Enzymatic reduction of "biogenic" aldehydes in brain. *Mol. Pharmacol.* **9**: 428-437.

THOMPSON, M.A. and WEINSHILBOUM, R.M. (1998). Rabbit lung indolethylamine *N*-methyltransferase. *J. Biol. Chem.* **273**: 34502-34510.

THOMPSON, M.A., MOON, E., KIM, U.-J., XU, J., SICILIANO, M.J. and WEINSHILBOUM, R.M. (1999). Human indolethylamine *N*-methyltransferase: cDNA cloning and experssion, gene cloning, and chromosomal localization. *Genomics* **61**: 285-297.

TODAKA, T., ISHIDA, T., KITA, H., NARIMATSU, S. AND YAMANO, S. (2005). Bioactivation of morphine in human liver: Isolation and identification of morphinone, a toxic metabolite. *Biol. Pharm. Bull.* **28**: 1275-1280.

TODAKA, T., YAMANO, S. and TOKI, S. (2000). Purification and characterization of NAD-dependent morphine 6-dehydrogenase from hamster liver cytosol, a new member of the aldo-keto reductase superfamily. *Arch. Biochem. Biophys.* **374**: 189-197.

TURNER, A.J., BAKER, K.M., ALGERI, S., FRIGERIO, A. and GARATTINI, S. (1974). Tetrahydropapaveroline: Formation *in vivo* and *in vitro* in rat brain. *Life Sci.* **14**: 2247-2257.

WEITZ, C.J., FAULL, K.F. and GOLDSTEIN, A. (1987). Synthesis of the skeleton of the morphine molecule by mammalian liver. *Nature* **330**: 674-677.

WEITZ, C.J., LOWNEY, L.I., FAULL, K.F., FEISTNER, G. and GOLDSTEIN, A. (1986). Morphine and codeine from mammalian brain. *Proc. Natl. Acad. Sci. USA* **83**: 9784-9788.

WILLIAMS, P.A., COSME, J., VINKOVIC, D.M., WARD, A., ANGOVE, H.C., DAY, P.J., VONRHEIN, C., TICKLE, I.J. and JHOTI, H. (2004). Crystal structures of human cytochrome P450 3A4 bound to metyrapone and progesterone. *Science* **305**: 683-686.

WINTERSTEIN, E. and TRIER, G. (1910). Die Alkaloide. Gebrüder Bornträger, Berlin.

XU, B.Q., AASMUNDSTAD, T.A., CHRISTOPHERSEN, A.S., MØRLAND, J. and BJØRNEBOE, A. (1997). Evidence for CYP 2D1-mediated primary and secondary O-dealkylation of ethylmorphine and codeine in rat liver microsomes. *Biochem. Pharmacol.* **53**: 603-609.

YAMAMOTO, Y., TASAKI, T., NAKAMURA, A., IWATA, H., KAZUSAKA, A., GONZALEZ, F.J. and FUJITA, S. (1998). Molecular basis of the Dark Agouti rat drug oxidation polymorphism: importance of CYP 2D1 and CYP 2D2. *Pharmacogenetics* **8**: 73-82.

YAMANO, S., KAGEURA, E., ISHIDA, T. and TOKI, S. (1985). Purification and characterization of guinea pig liver morphine 6-dehydrogenase. *J. Biol. Chem.* **260**: 5259-5264.

YAMANO, S., TAKAHASHI, A., TODAKA, T. and TOKI, S. (1997). In vivo and in vitro formation of morphinone from morphine in rat. *Xenobiotica* **27**: 645-656.

YAMAZAKI, H., NAKAMURA, M., KOMATSU, T., OHYAMA, K., HATANAKA, N., ASAHI, S., SHIMADA, N., GUENGERICH, F.P., SHIMADA, T., NAKAJIMA, M. and YOKOI, T. (2002). Roles of NADPH-P450 reductase and apo- and holo-cytochrome b_5 on xenobiotic oxidations catalyzed by 12 recombinant human cytochrome P450s expressed in membranes of *Escherichia coli*. *Protein Expres. Purif.* **24**: 329-337.

YAMAZAKI, S., SATO, K., SUHARA, K., SAKAGUCHI, M., MIHARA, K. and OMURA, T. (1993). Importance of the proline-rich region following signal-anchor sequence in the formation of correct conformation of microsomal cytochrome P-450s. *J. Biochem.* **114**: 652-657.

YEH, S.H., GORODETZKY, C.W. and KREBS, H.A. (1977). Isolation and identification of morphine 3- and 6-glucuronides, morphine 3,6-diglucuronide, morphine 3-ehteral sulfate, normorphine, and normorphine 6-glucuronide as morphine metabolites in humans. *J. Pharm. Sci.* **66**: 1288-1293.

YOSHIOKA, S., TAKAHASHI, S., HORI, H., ISHIMORI, K. and MORISHIMA, I. (2001). Proximal cysteine residue is essential for the enzymatic activities of cytochrome $P450_{cam}$. *Eur. J. Biochem.* **268**: 252-259.

YU, A., KNELLER, B.M., RETTIE, A.E. and HAINING, R.L. (2002). Expression, purification, biochemical characterization, and comparative functions of human P450 2D6.1, 2D6.2, 2D6.10, and 2D6.17 allelic isoforms. *J. Pharmacol. Exp. Ther.* **303**: 1291-1300.

YUE, Q.Y. and SÄWE, J. (1997). Different effects of inhibitors on the O- and N-demethylation of codeine in human liver microsomes. *Eur. J. Clin. Pharmacol.* **52**: 41-47.

ZHANG, J., FERGUSON, S.S., BARAK, L.S., BODDULURI, S.R., LAPORTE, S.A., LAW, P.Y. and CARON, M.G. (1998). Role for G protein-coupled receptor kinase in agonist-specific regulation of μ-opioid receptor responsiveness. *Proc. Natl. Acad. Sci. USA* **95**: 7157-7162.

ZHU, W., BAGGERMAN, G., GOUMON, Y., CASARES, F., BROWNAWELL, B. and STEFANO, G.B. (2001a). Presence of morphine and morphine-6-glucuronide in the marine mollusk *Mytilus edulis* ganglia determined by GC/MS and Q-TOF-MS. Starvation increases opiate alkaloid levels. *Mol. Brain Res.* **88**: 155-160.

ZHU, W., BILFINGER, T.V., BAGGERMAN G., GOUMON, Y. AND STEFANO, G.B. (2001b). Presence of endogenous morphine and morphine 6 glucuronide in human heart tissue. *Int. J. Mol. Med.* **7**: 419-422.

ZHU, W., CADET, P., BAGGERMAN, G., MANTIONE, K. and STEFANO, G.B. (2005). Human white blood cells synthesize morphine: CYP 2D6 modulation. *J. Immunol.* **175**: 7357-7362.

ZIEGLER, J., VOIGTLÄNDER, S., SCHMIDT, J., KRAMELL, R., MIERSCH, O., AMMER, C., GESELL, A. and KUTCHAN, T.M. (2006). Comparative transcript and alkaloid profiling in *Papaver* species identifies a short chain dehydrogenase/reductase involved in morphine biosynthesis. *Plant J.* **48**: 177-192.

Acknowledgement

I wish to express my deep gratitude to Prof. M.H. Zenk for giving me the opportunity to work in this fascinating and challenging field, for his consistent interest in the progress of the studies, for his sincerity and care, for his extraordinary advices, encouragement and support during the development of the dissertation and for teaching me the important aspects of research and life.

My sincere appreciation is also to Prof. T.M. Kutchan for her scientific and personal advices and for her continuous support over the past years.

I would like to thank Prof. B. Dräger for her constant and invaluable support, organization and contribution to accomplish this dissertation.

My special thanks go to Prof. M. Spiteller and Dr. M. Lamshöft for the realization and interpretation of mass spectrometric analyses and for the excellent scientific collaboration. I would also like to acknowledge Prof. M. Spiteller for the generous opportunity of a short stay in his research group and the outstanding hospitality.

I would like to thank Dr. L. Hicks and Dr. B. Zhang for introducing me into mass spectrometry and for their guidance and constant support.

I would like to offer my appreciation Prof. F.P. Guengerich for the generous gift of human CYP 2D6 and rat cytochrome P450 reductase, for great collaboration and for valuable discussion.

I wish to thank the entire Animal Facility at Washington University for the welcoming and friendly work environment and, particularly, Dr. VMD T. Keadle for her effort, generosity and kindness.

I deeply appreciated the efforts of Mrs. J. Coyle, especially for her linguistic help.

Atlas of High-Resolution Mass Spectra (LTQ-Orbitrap)

Full scan:

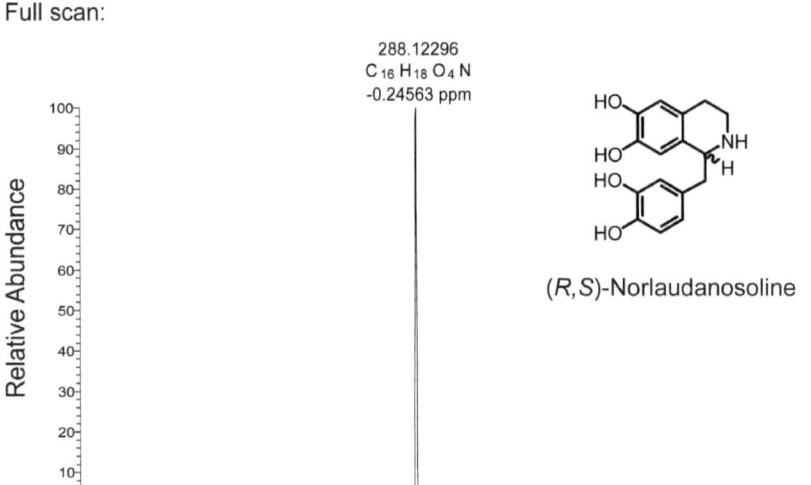

288.12296
$C_{16}H_{18}O_4N$
-0.24563 ppm

(R,S)-Norlaudanosoline

MS/MS:

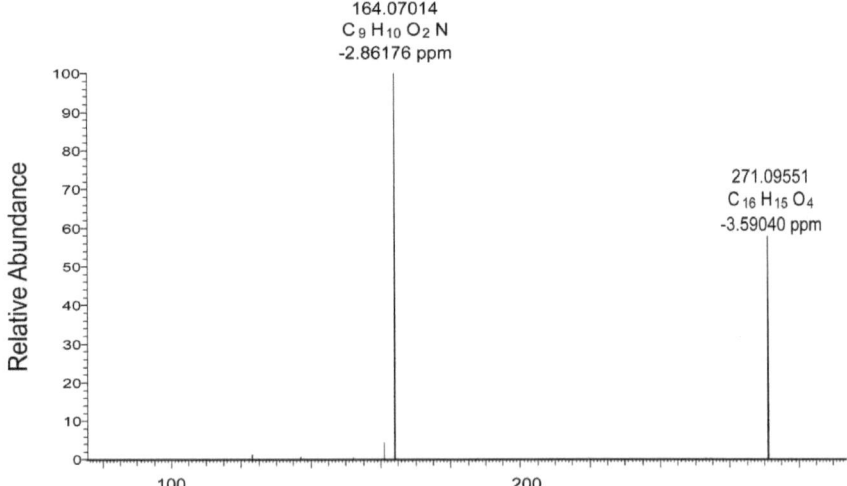

164.07014
$C_9H_{10}O_2N$
-2.86176 ppm

271.09551
$C_{16}H_{15}O_4$
-3.59040 ppm

Full Scan:

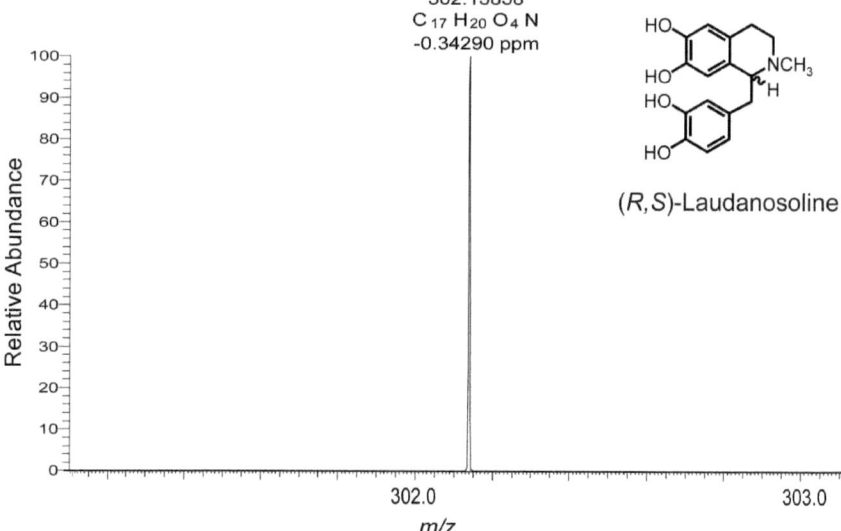

MS/MS:

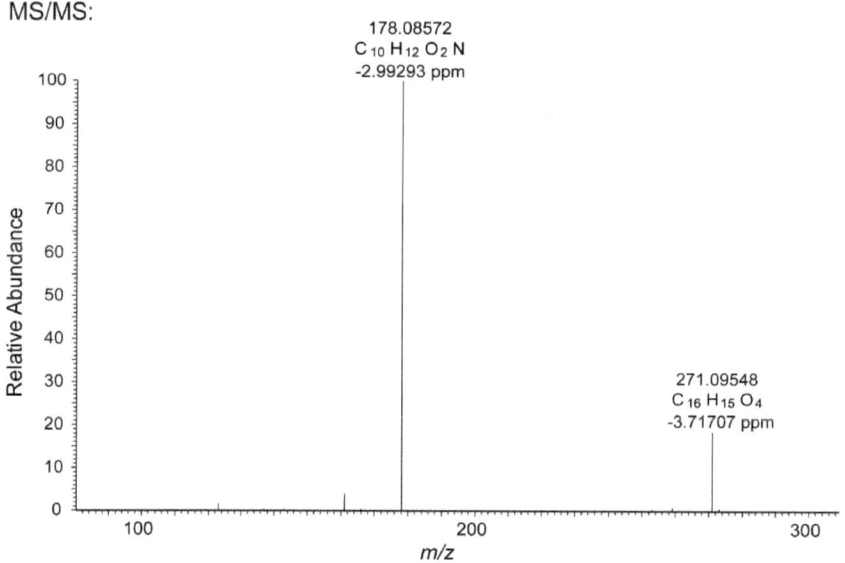

Full Scan:

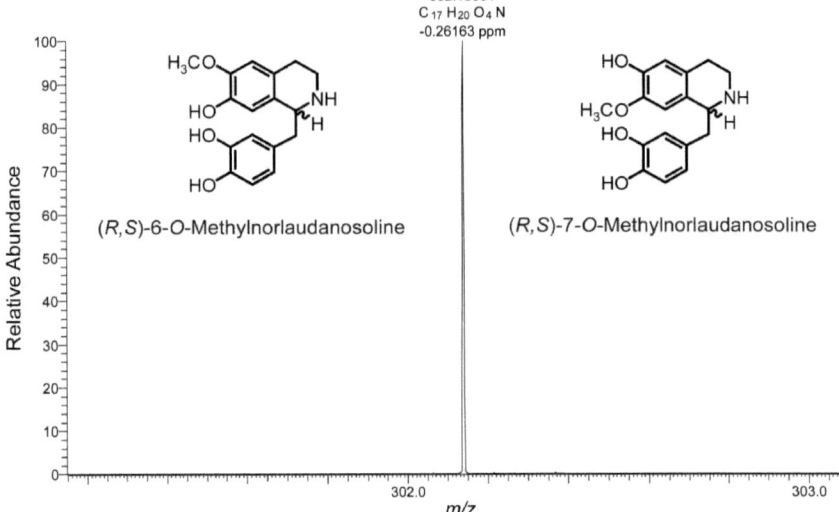

MS/MS:

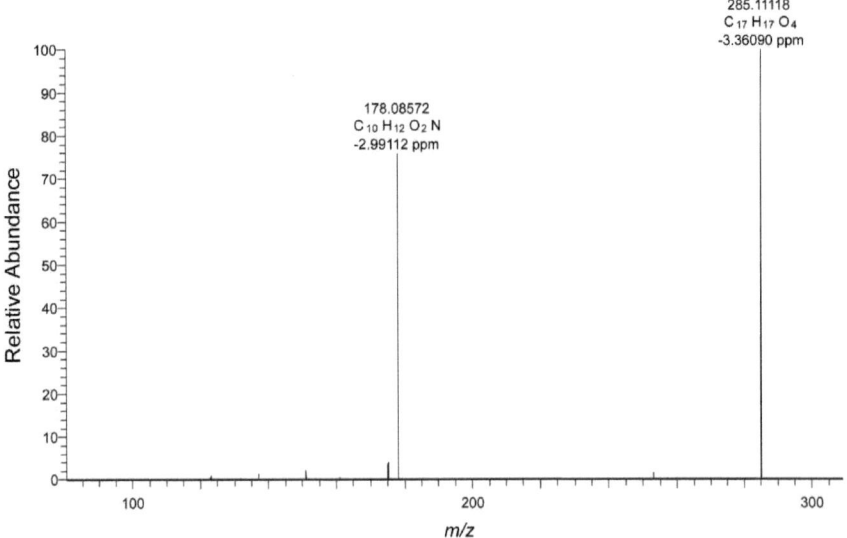

Full Scan:

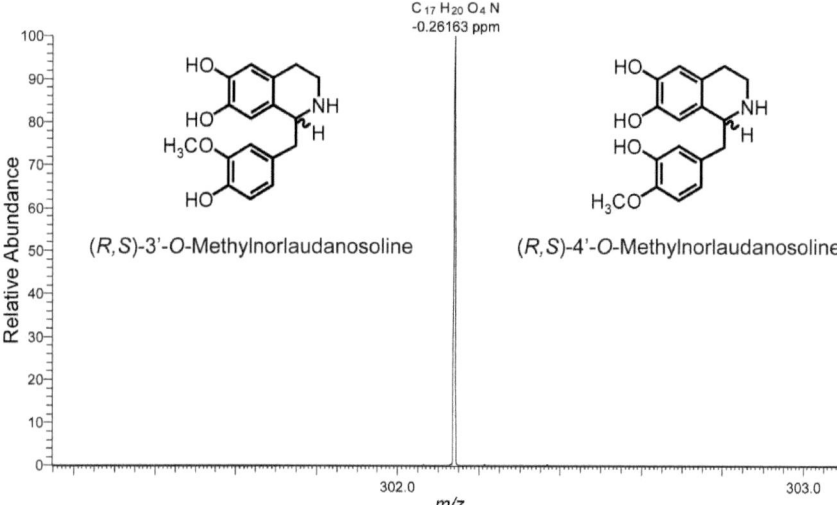

MS/MS:

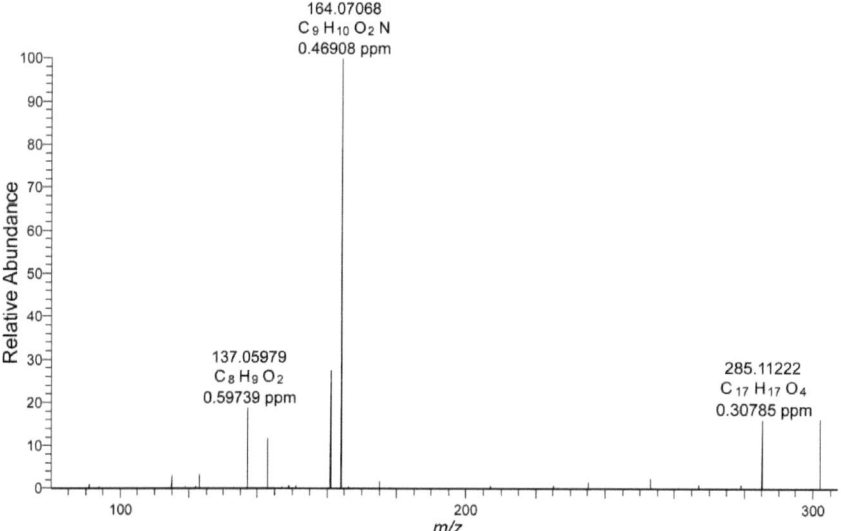

Full scan:

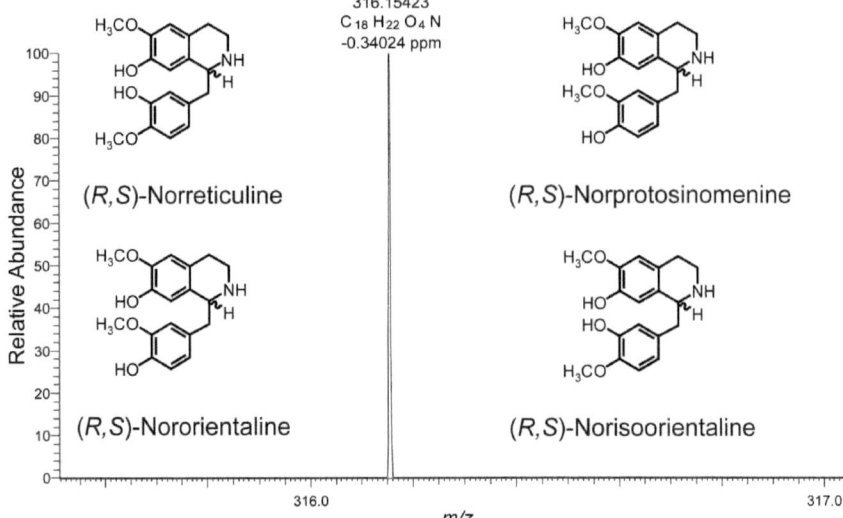

MS/MS:

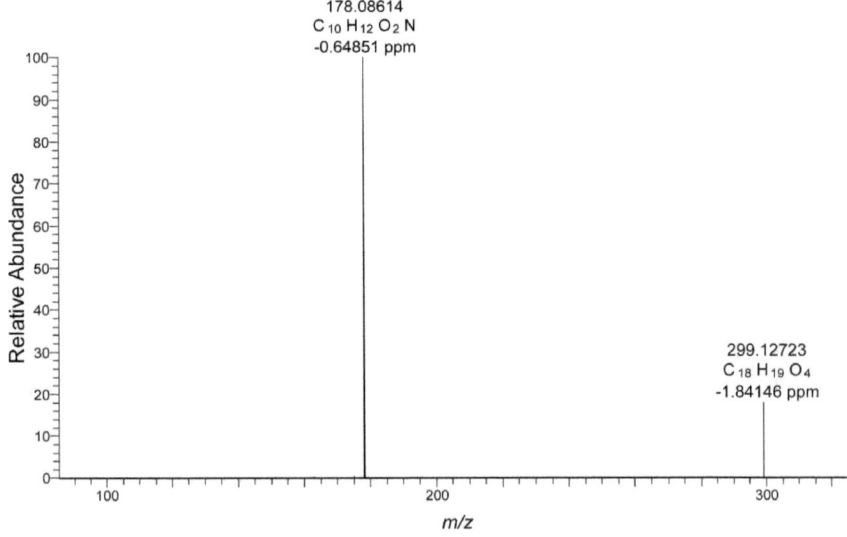

Full scan:

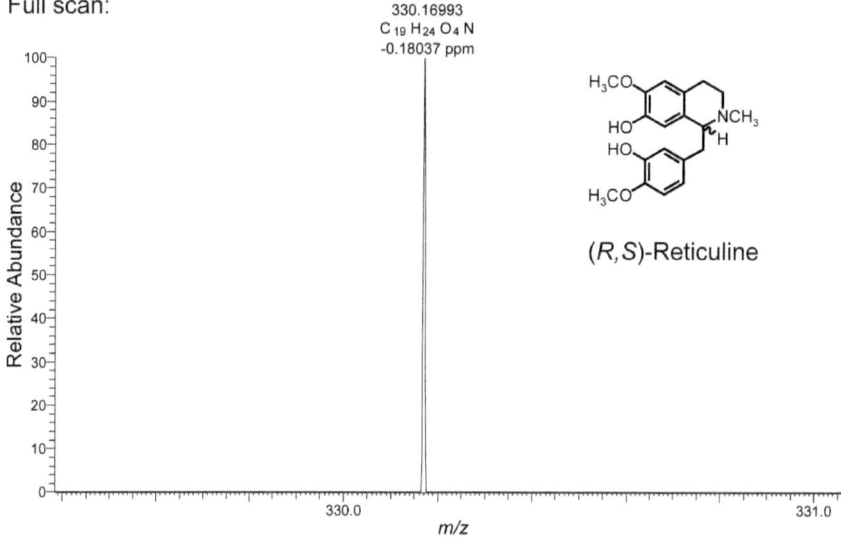

MS/MS:

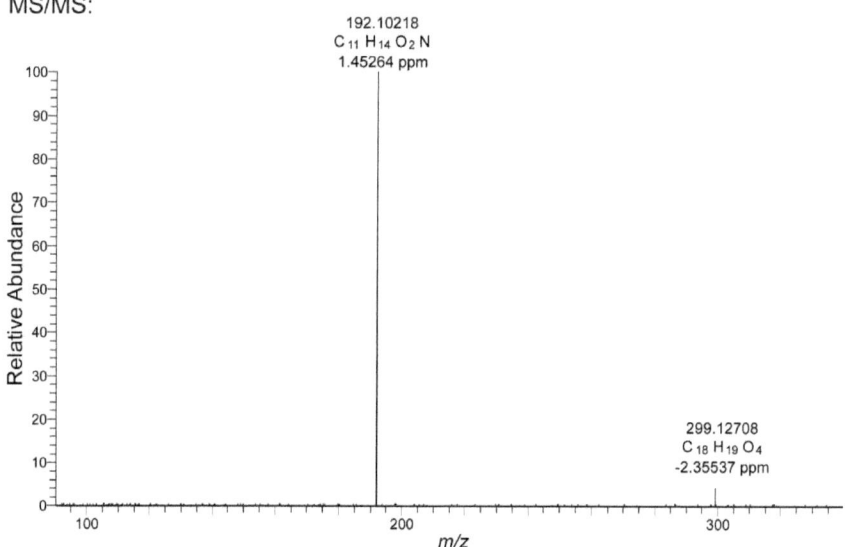

Full scan:

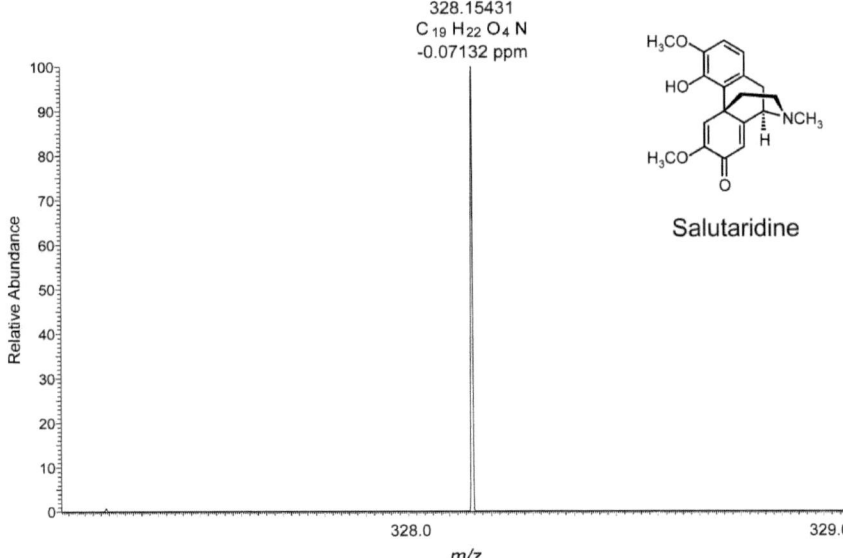

MS/MS:

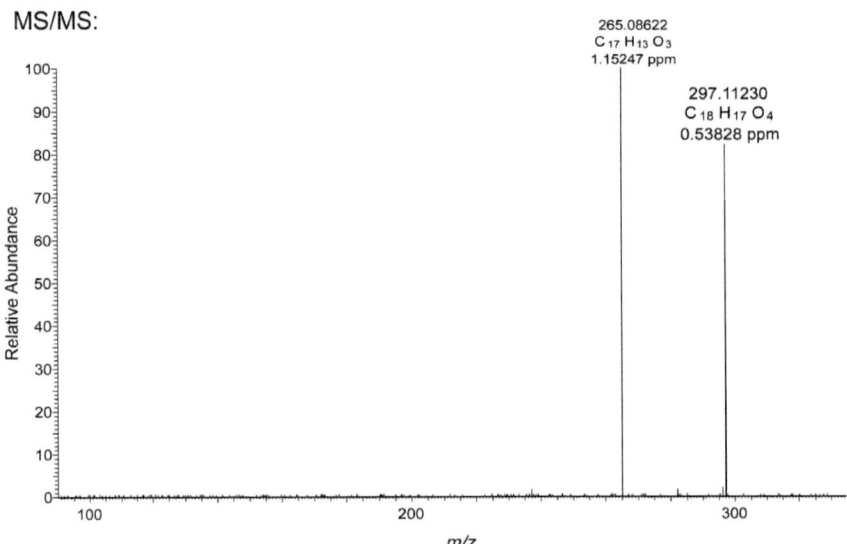

Full scan:

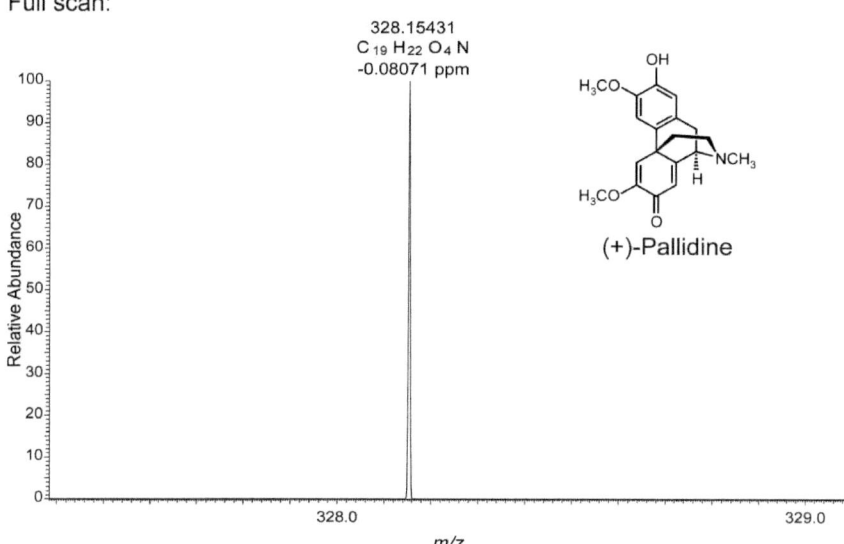

MS/MS:

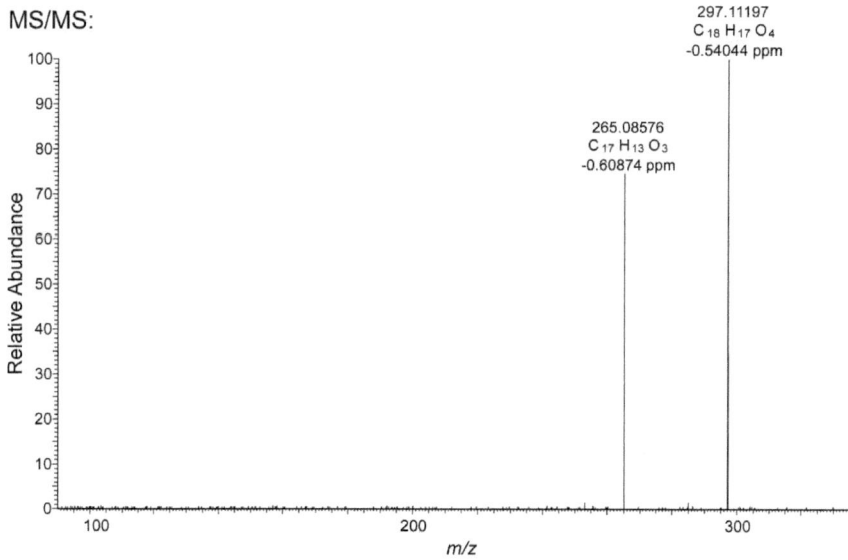

Full scan:

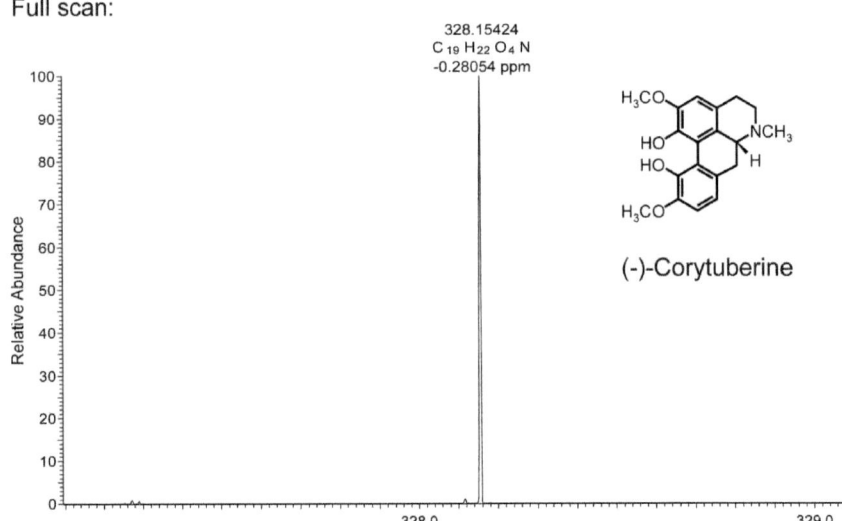

MS/MS:

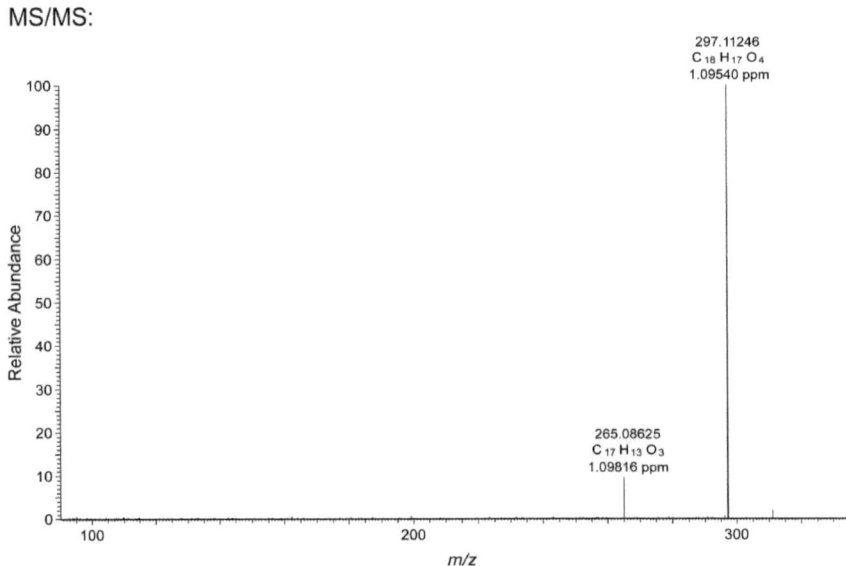

Full scan:

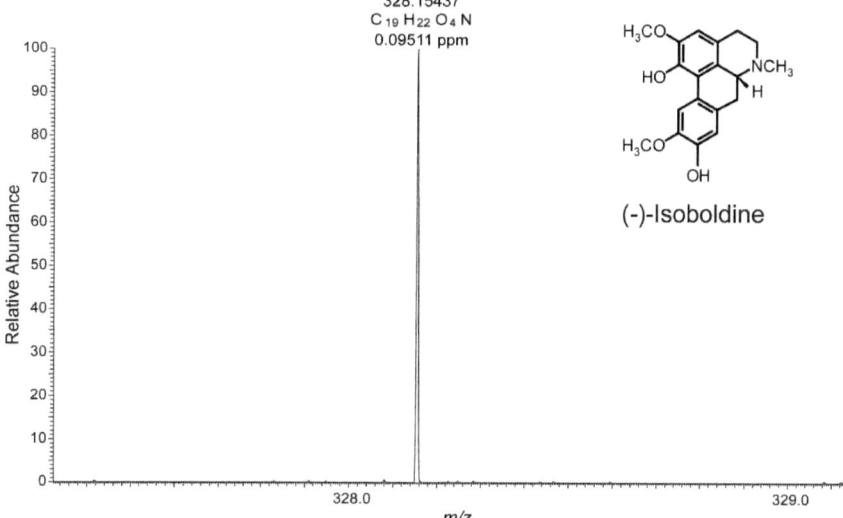

MS/MS:

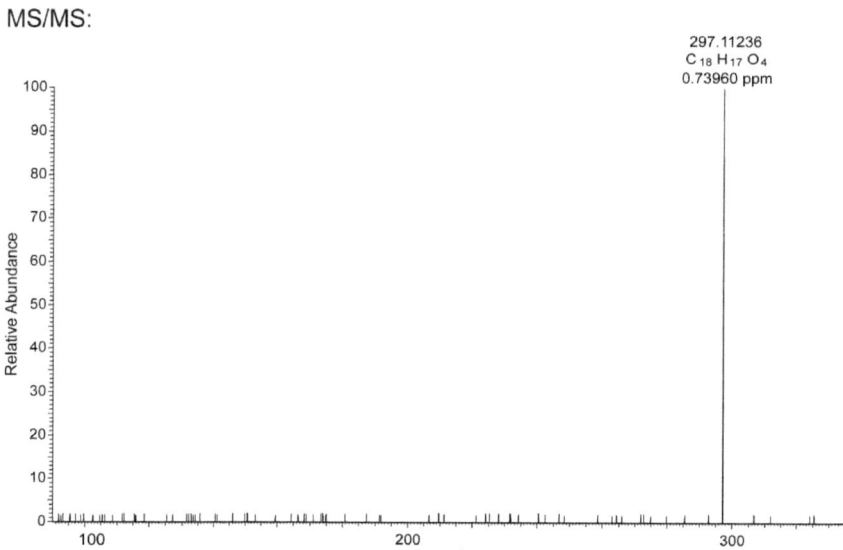

Full scan:

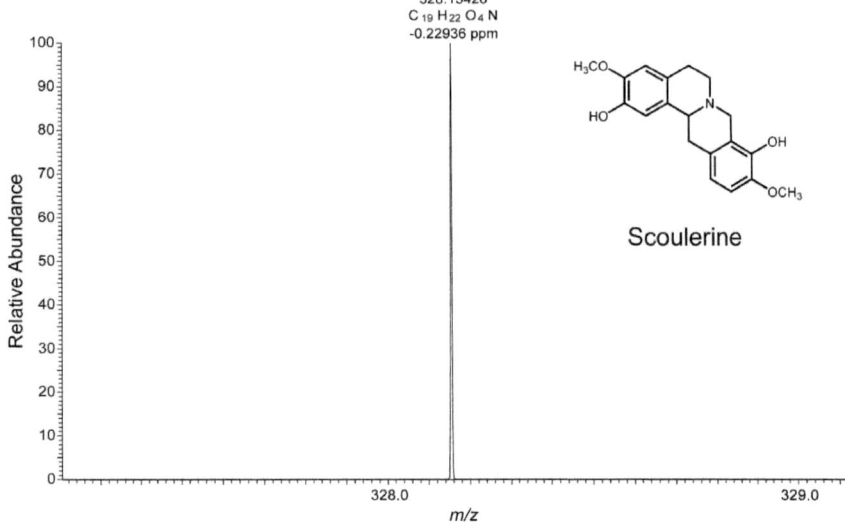

MS/MS:

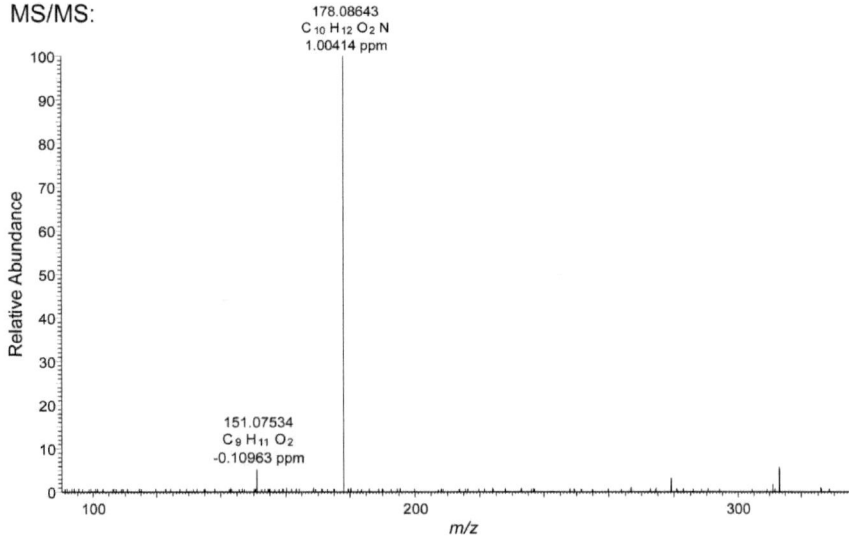

Full scan:

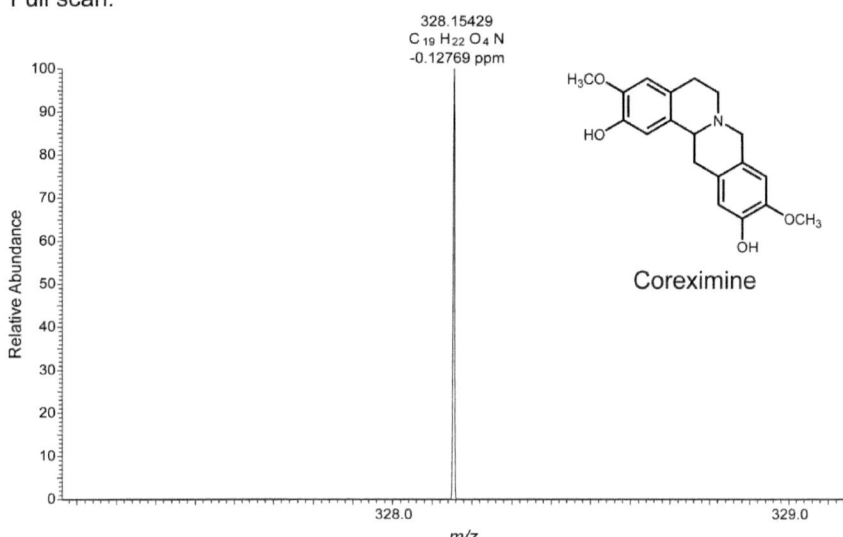

MS/MS:

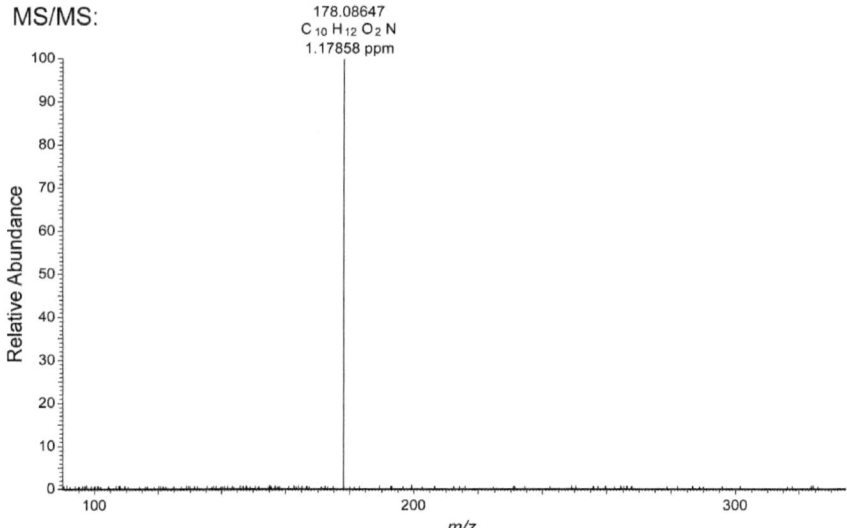

Full scan:

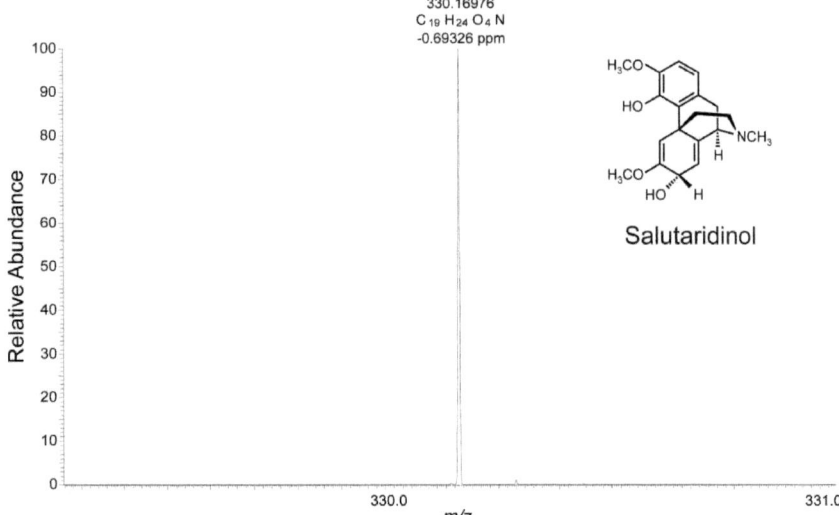

MS/MS:

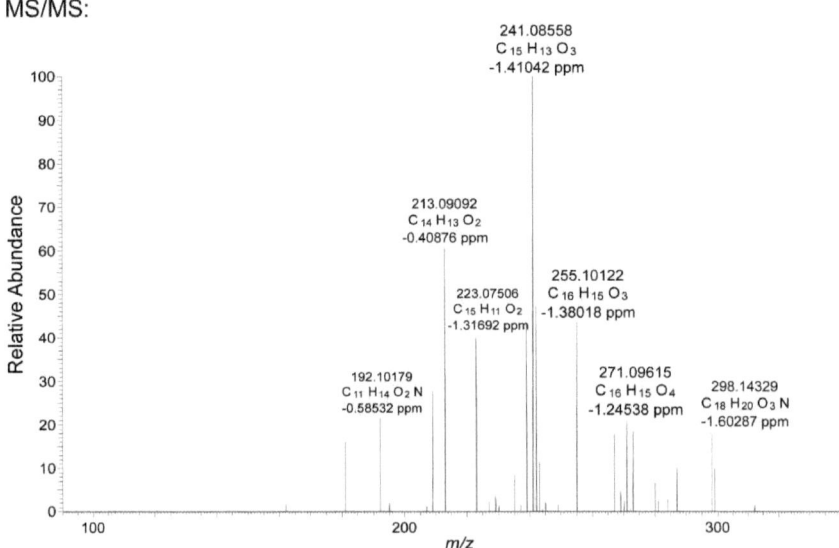

Full scan:

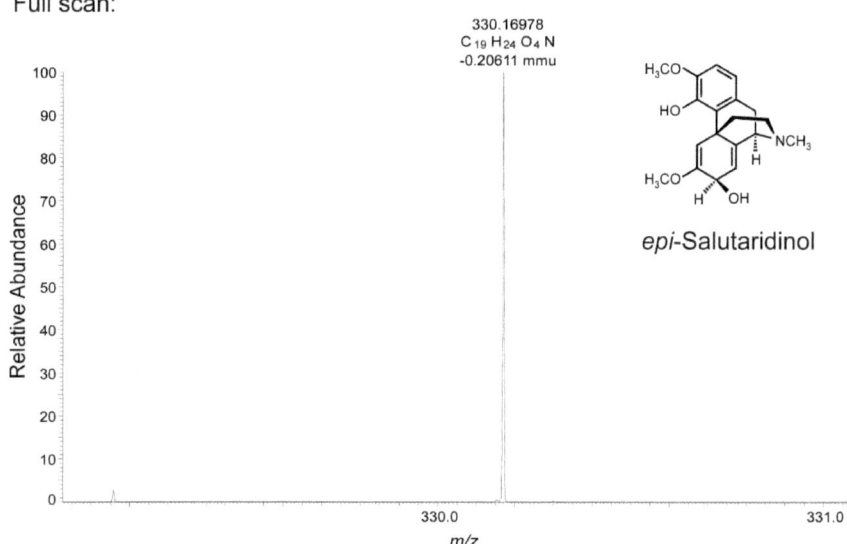

epi-Salutaridinol

MS/MS:

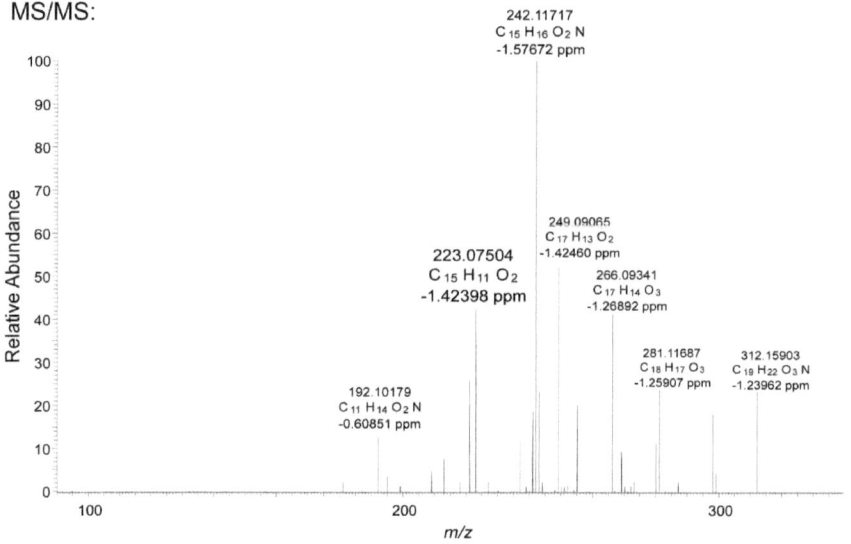

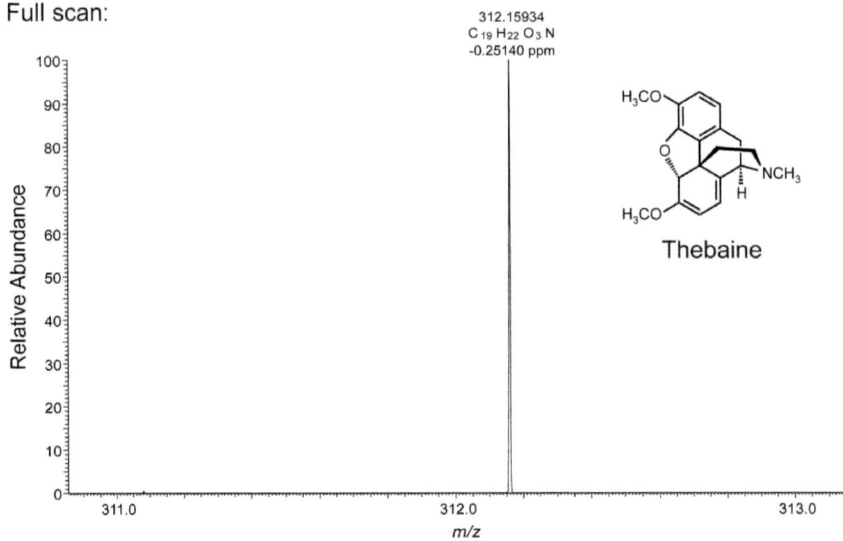

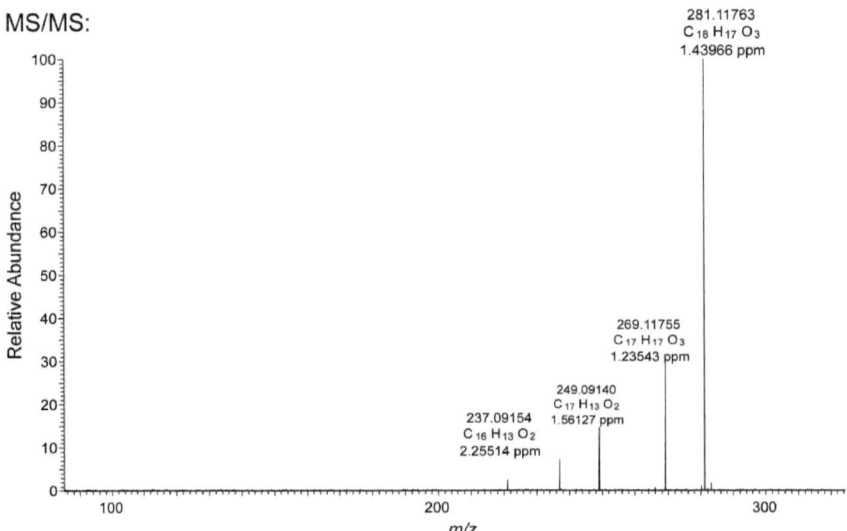

Full scan:

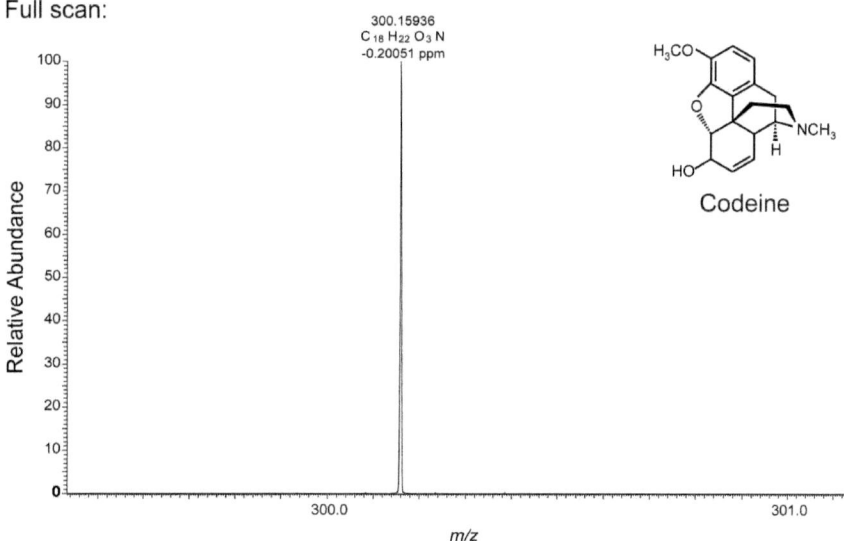

MS/MS:

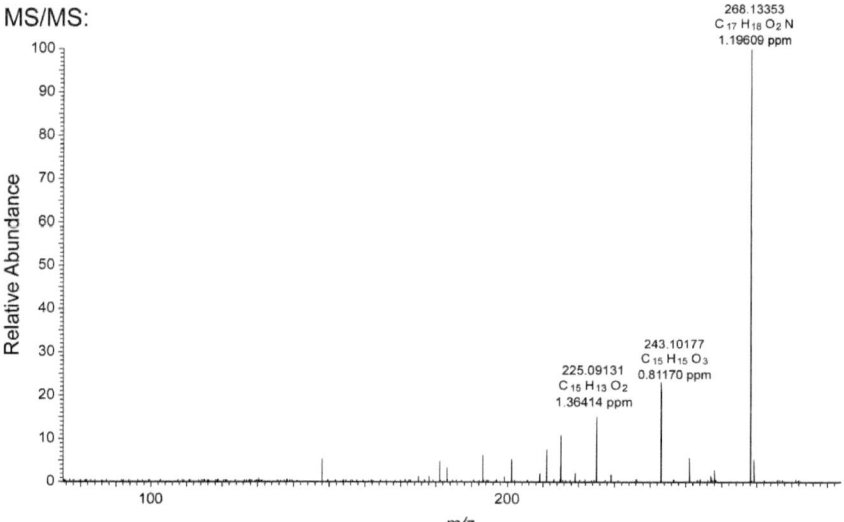

Full scan:

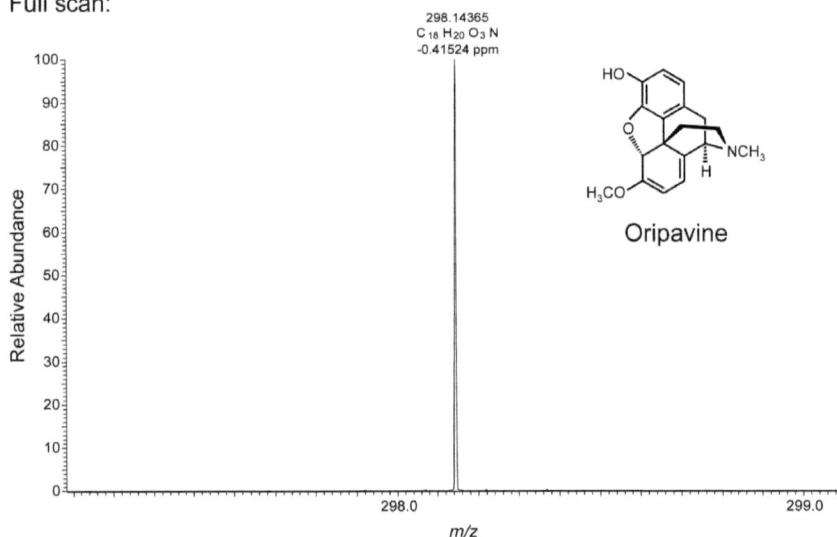

MS/MS:

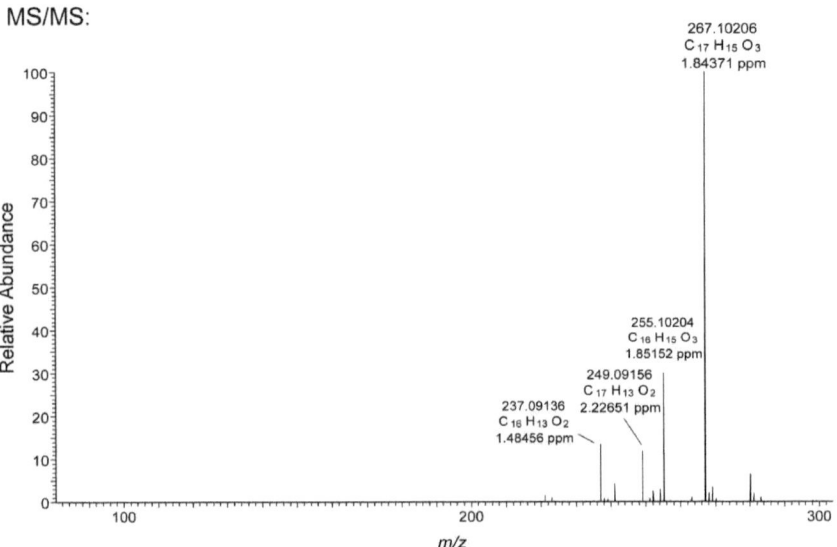

Full scan:

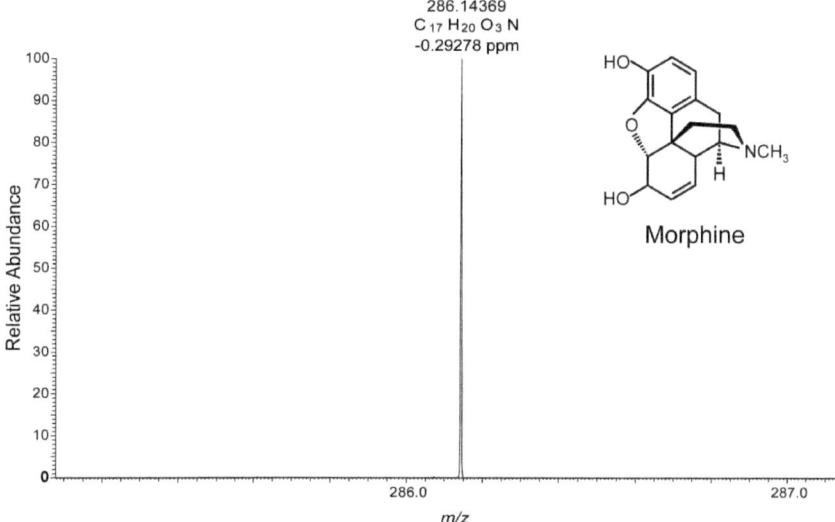

MS/MS:

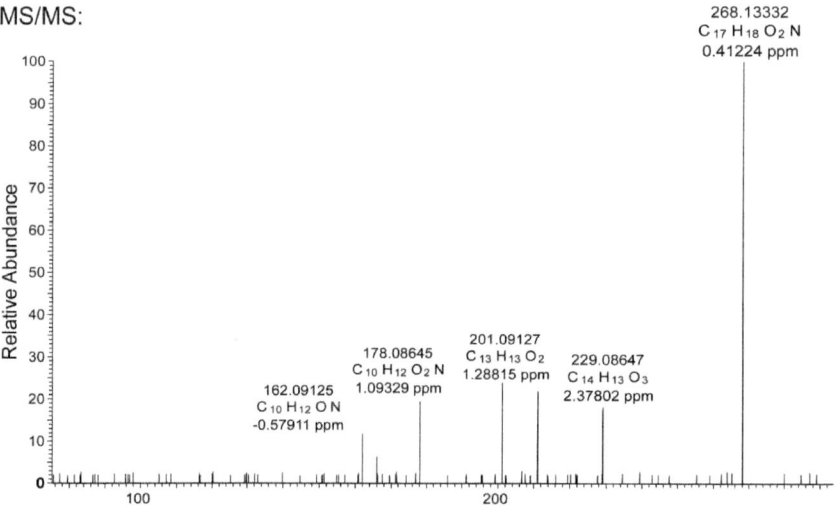

i want morebooks!

Buy your books fast and straightforward online - at one of world's fastest growing online book stores! Environmentally sound due to Print-on-Demand technologies.

Buy your books online at
www.get-morebooks.com

Kaufen Sie Ihre Bücher schnell und unkompliziert online – auf einer der am schnellsten wachsenden Buchhandelsplattformen weltweit! Dank Print-On-Demand umwelt- und ressourcenschonend produziert.

Bücher schneller online kaufen
www.morebooks.de

VDM Verlagsservicegesellschaft mbH
Heinrich-Böcking-Str. 6-8   Telefon: +49 681 3720 174   info@vdm-vsg.de
D - 66121 Saarbrücken   Telefax: +49 681 3720 1749 www.vdm-vsg.de

Printed by Books on Demand GmbH, Norderstedt / Germany